GETTING IN

THE ESSENTIAL GUIDE TO FINDING A STEMM UNDERGRAD RESEARCH EXPERIENCE

PARIS H. GREY &
DAVID G. OPPENHEIMER

Second Edition

The University of Chicago Press
Chicago and London

The University of Chicago Press, Chicago 60637
The University of Chicago Press, Ltd., London

Published 2023
Printed in the United States of America

32 31 30 29 28 27 26 25 24 23   1 2 3 4 5

ISBN-13: 978-0-226-82520-5 (cloth)
ISBN-13: 978-0-226-82541-0 (paper)
ISBN-13: 978-0-226-82540-3 (e-book)
DOI: https://doi.org/10.7208/chicago/9780226825403.001.0001

Originally published in 2015 by Secret Handshake Press. Any questions concerning permissions should be directed to the Permissions Department at the University of Chicago Press, Chicago, IL.

Library of Congress Cataloging-in-Publication Data

Names: Grey, Paris H., author. | Oppenheimer, David G., author.
Title: Getting in : the essential guide to finding a STEMM undergrad research experience / Paris H. Grey and David G. Oppenheimer.
Other titles: Chicago guides to academic life.
Description: Second edition. | Chicago : The University of Chicago Press, 2023. | Series: Chicago guides to academic life | Includes bibliographical references and index.
Identifiers: LCCN 2022049579 | ISBN 9780226825205 (cloth) | ISBN 9780226825410 (paperback) | ISBN 9780226825403 (e-book)
Subjects: LCSH: College students—Research. | Science—Study and teaching (Higher). | Research.
Classification: LCC Q181.G7659 2023 | DDC 507.1/1—dc23/eng20230117
LC record available at https://lccn.loc.gov/2022049579

♾ This paper meets the requirements of ANSI/NISO Z39.48-1992 (Permanence of Paper).

GETTING IN

CHICAGO GUIDES TO ACADEMIC LIFE

A STUDENT'S GUIDE TO LAW SCHOOL
Andrew B. Ayers

WHAT EVERY SCIENCE STUDENT SHOULD KNOW
Justin L. Bauer

THE CHICAGO GUIDE TO YOUR CAREER IN SCIENCE
Victor A. Bloomfield and Esam El-Fakahany

THE CHICAGO HANDBOOK FOR TEACHERS, SECOND EDITION
Alan Brinkley, Esam El-Fakahany, Betty Dessants, Michael Flamm, Charles B. Forcey Jr., Matthew L. Ouellett, and Eric Rothschild

THE CHICAGO GUIDE TO LANDING A JOB IN ACADEMIC BIOLOGY
C. Ray Chandler, Lorne M. Wolfe, and Daniel E. L. Promislow

THE PHDICTIONARY
Herb Childress

BEHIND THE ACADEMIC CURTAIN
Frank F. Furstenberg

THE CHICAGO GUIDE TO YOUR ACADEMIC CAREER
John A. Goldsmith, John Komlos, and Penny Schine Gold

HOW TO SUCCEED IN COLLEGE (WHILE REALLY TRYING)
Jon B. Gould

LIFE AND RESEARCH
Paris H. Grey and David G. Oppenheimer

57 WAYS TO SCREW UP IN GRAD SCHOOL
Kevin D. Haggerty and Aaron Doyle

HOW TO STUDY
Arthur W. Kornhauser

DOING HONEST WORK IN COLLEGE
Charles Lipson

SUCCEEDING AS AN INTERNATIONAL STUDENT IN THE UNITED STATES AND CANADA
Charles Lipson

OFF TO COLLEGE
Roger H. Martin

A complete list of series titles is available on the University of Chicago Press website.

For Lilli and the rest of the warren.
And for Tree—you know why.
PHG

For my parents, Glen and Betty.
And for my brothers, Alan and Dale.
DGO

CONTENTS

PREFACE

Getting In is a companion for the beginning of your research journey. It's a guide to finding a meaningful undergraduate research experience and tips for preparing for your research adventure. *Getting In* is for all students interested in a research experience, whether your goal is using research as a career launchpad or you're simply interested in discovering something new (although maybe you have both goals!). *Getting In* can also fill in some of the gaps shortly after joining a research group, or if you're considering a new opportunity because your current one doesn't meet your expectations.

If you're thinking, "Why would I need a *book* to find an undergrad research position?" the answer is straightforward: to help you successfully navigate the hidden curriculum of finding a research *experience* that aligns with your goals and is a meaningful use of your time. And as you implement strategies during your search, you'll develop skills and habits you'll use to build a productive, professional relationship with your research mentor when you start conducting research.

WHAT IS THE HIDDEN CURRICULUM?

The term *hidden curriculum* refers to the unofficial lessons that are presumed to be known by a student simply because they went to high school or attend college. However, these practices aren't taught widely or formally in a classroom, like lecture material, but instead are invisible or obscured. Even students familiar with the broader *college* hidden curriculum often aren't aware of the expectations specific to an undergrad research experience. But this knowledge gap is especially penalizing for students from underserved and underrepresented communities who don't have access to customized mentorship early in their undergrad studies.

Our definition of students from underrepresented communities starts with groups identified by the United States government agencies the

National Institutes of Health and National Science Foundation (NSF). When we wrote this edition, those agencies identified individuals from groups as underrepresented racial and ethnic groups in biomedical, clinical, behavioral, and social sciences as people who identify as Black or African American, Hispanic or Latino, American Indian or Alaska Native, Native Hawaiian, and other Pacific Islanders.[1] The NSF also includes in this identification of underrepresented groups individuals who manage disabilities or a chronic health matter, and people from low socioeconomic backgrounds. In *Getting In*, we expand these identities to include LGBTQIA+ students, neurodivergent students, first-generation college students, and students who are United States military veterans.[2] We also recognize that someone's whole identity may include them in several of these communities. Because the hidden curriculum disproportionately affects students from underserved communities, the associated practices contribute to inequity and the maintaining of nondiverse, noninclusive spaces in science.

For undergrad research specifically, the hidden curriculum includes knowing the potential interpersonal and professional benefits of conducting research, expectations some professors and administrators have for students interacting with them through email or in person, understanding that there are several types of research group cultures, knowing that research failures and challenges are normal occurrences during the learning process and are unreliable metrics to gauge future success, and more. In *Getting In* we aim to demystify some parts of the hidden curriculum connected to undergrad research in science, technology, engineering, math, and medicine (STEMM) and provide actionable strategies to navigate these informal rules and practices.

For example, common advice given to an undergrad in search of a research opportunity is to either read several scientific articles published by a professor or check online campus databases for labs with open positions. Then, regardless of which method is selected, contact the professor, or someone else in the research group, while demonstrating enthusiasm for the study and explaining how you will contribute to the research program. We admit that over the years we gave similar advice when a student asked, "How do I find a research position?" until we wondered why, if the process

1. National Institutes of Health, "Populations Underrepresented in the Extramural Scientific Workforce," updated March 30, 2022, https://diversity.nih.gov/about-us/population-underrepresented.

2. We encourage all readers to familiarize themselves with resources for LGBTQIA+ and neurodiverse individuals. Although there are numerous resources to do this, two we recommend are GLAAD (https://www.glaad.org/) and the Stanford Neurodiversity Project (https://med.stanford.edu/neurodiversity.html).

was typically this streamlined, so many students were still asking for help *every* semester. So, being scientists, we set out to evaluate the advice by connecting with some of the current and former undergrads from our lab and various undergrad research mentors.

We learned, among other things, that giving this well-intentioned advice in the absence of guidance on how to implement it can quickly discourage first-generation college students, students who have never knowingly interacted with a researcher who shares their identities, and students on campuses with limited research opportunities. Essentially, each method has potential pitfalls that can discourage an otherwise enthusiastic and motivated undergrad from starting the search for a research experience or cause them to give up early in the process. But we also know that reading a scientific paper or using an online, on-campus database works for some undergrads, so we agree that these are valuable strategies for those students. Therefore, in *Getting In*, we discuss how to maximize these approaches, but we also cover potential issues of each approach and recommend alternative search methods for undergrads who need or want them.

WHAT IS A MEANINGFUL RESEARCH EXPERIENCE?

Even after you've successfully navigated the search and application processes, it's important to consider whether an opportunity is likely to be a meaningful use of your time. Our definition of a meaningful research experience has two components: (1) it aligns with your values, expectations, and goals—provided those expectations and goals are realistic; and (2) it provides opportunities to actively conduct research in a lab, clinic, field site, or other environment so you're not solely performing research-adjacent tasks or lab-keeping chores. We provide interview questions you'll ask to understand the realities of an opportunity and a potential mentor's expectations. Together, the answers will help you (and your mentor) avoid frustration and disappointment related to some common misunderstandings down the road.

HOW WE ORGANIZED *GETTING IN*

Whether you attend an institution with limited or numerous opportunities, undergrad research positions are competitive. *Getting In* is your insiders' guide to demystifying some aspects of undergrad research in STEMM and empowering you to navigate the hidden curriculum. With these goals in mind, we organized *Getting In* into two parts. Part 1, comprising chapters 1–3, includes potential benefits for conducting undergrad research, a primer on some types of STEMM research, an overview of how various research groups function to build a lab culture, and tips for identifying your goals and managing your expectations for a research experience. Part 2, comprising chapters 4–6, covers detailed strategies on how to search, apply, and interview for a research position that is compatible with the time you have available and the goals you want to achieve through it. But if you're worried about the length of *Getting In*, we want to reassure you: we know that you're busy and sometimes will be short on time. So, although all the information in *Getting In* is important, we strategically incorporated headers, lists, and **boldface** text to help you learn the **essential information** or to find it again when you need to quickly review a section.

WHERE TO FIND TIPS ON CONDUCTING RESEARCH

In *Getting In*, we don't explain procedures for conducting research after joining a research group or provide detailed strategies for navigating the daily hidden curriculum that accompanies working in a lab environment. This is because best practices on these topics are too complex to cover meaningfully in this text. When you begin searching for a research experience, we recommend following us on our socials: on Twitter @YouInTheLab (Undergrad in the Lab), and on Facebook and Instagram @UndergradInTheLab for tips and articles that we share through these channels from a variety of sources. We also recommend our website UndergradInTheLab.com. Although many of the articles are for students involved in wet lab research (that is, in a space equipped with liquid chemicals and reagents) in the biological sciences such as cell biology, microbiology, chemistry, and similar disciplines, some content is relevant to all undergrads regardless of their research field. When we publish our next book for undergrads conducting research in a professional lab, we'll announce it in those spaces.

For students in their final year of undergrad study and preparing to attend graduate school or a postbaccalaureate program, applying for research jobs after graduation, or planning to conduct research as part of a medical school program, we recommend another book we wrote, *Life and Research: A Survival Guide for Early-Career Biomedical Scientists*, published by the University of Chicago Press. This guidebook is for early-career researchers and contains strategies on how to identify and develop underutilized skills, ward off common problems with labmates by keeping the lines of communication open, and build a network of mentors.

For resources on your campus, we recommend checking in with the Office of Undergraduate Research (if you have one) for programs or seminars on research-related topics such as how to present your research at a conference, write an abstract, apply for research-related scholarships, and find other workshops or presentations. Also, check the course catalog for your home department for classes that teach beginning or advanced research methods and topics related to science disciplines in your major. If you're interested in writing a senior thesis or submitting your work for publication in an undergrad research journal, connect with the professionals at the writing or Career Resource Center early on for help on creating an outline and ensuring that the format is in line with your college's or a journal's expectations. In addition, an academic advisor, a preprofessional advisor, or a professor you already consider to be mentor can be sources for advice and guidance on research-related topics as well as other college or personal matters. And we strongly encourage you to regularly connect with your research mentor and labmates!

A NOTE ON OUR USE OF EVERYDAY INCLUSIVE LANGUAGE

To make the text of this book gender inclusive and gender neutral, we selected *they*, *them*, *their*, and *theirs* to represent a singular person. We also avoided ableist language, although we acknowledge that in some instances other writers with the same goal might make a different word choice. For example, we chose "a student managing a disability" as opposed to "a disabled student" when applicable. But we know that an individual might choose language-first or disability-first terminology to refer to themself. In person, we would respect the preference of a student based on how they describe themself. Although language evolves, we hope that the word choices we made in this edition will age well. Even so, we will continue our education on best practices for using inclusive language in everyday situations.

SHARING OUR AND OTHERS' PERSPECTIVES

Although we each bring a different perspective to *Getting In*—one from the lab mentor's bench and one from the professor's desk—to distinguish an individual experience from our collective one, we use our initials when needed. When the experience originates from Paris H. Grey, the initials are PHG, and when from David G. Oppenheimer, they are DGO. We also use opinions and reflections shared with us by colleagues and through our extensive connections on social media. Some of these influences have included undergrads, but we use the term *colleague* in *Getting In* for all who mentor undergrads in research, including professors, instructors, grad students, staff scientists, and administrative staff.

A NOTE TO MENTORS AND INSTRUCTORS FAMILIAR WITH THE FIRST EDITION

New material in this edition of *Getting In* includes the introduction of some topics and expansion of others previously covered in the first edition (published in 2015). We've added a new section in chapter 3 that covers summer undergrad research programs (SURPs), how to find them, and how to prepare for virtual interviews. In a new section titled "Qualifications for Participation in Undergrad Research," also in chapter 3, we address recurring concerns we've fielded from students in undergrad research interviews or on social media. We've expanded other material in chapter 3 to include additional discussions of expectations undergrads commonly have prior to starting research, such as how quickly they might earn a publication, the impact they can make through research, and how soon they'll be able to plan experiments. We also added information specifically for undergrads on a premedical path that busts the myth that they should misrepresent their career aspirations when applying to a research experience to increase the chance of receiving an offer to join a research group. And recognizing that many low-income students don't have the privilege to volunteer for a research position and instead need to receive a paycheck or credit that counts in their grade point average, we've expanded our advice about taking research for class credit, including weighing the potential financial consequences against the transcript advantages and searching for a paid position. Finally, we expanded the interview question section in chapter 6 to help students more fully understand their goals in relation to the available research opportunity and the difference between asking high-value and low-value questions. This distinction is intended to help undergrads discover whether their goals and expectations align with a potential mentor's expectations.

PART ONE

1

Why Choose Research?

WHAT'S IN IT FOR YOU?

Depending on your major, career path, and academic level, you may have already been advised, perhaps numerous times, that it's important to participate in undergraduate research.

If you're considering undergrad research, you probably already know some of the potential benefits: it can strengthen your applications for medical or graduate school, can help you explore a potential career path, and can lead to recommendation letters from the professor in whose group you conduct research. **We encourage you to participate in an undergrad research experience because the potential benefits go beyond the basics.** A meaningful undergrad research experience can also support your long-term professional development goals and help you gain crucial interpersonal skills that you'll need regardless of what your career path turns out to be.

Granted, if your career goals include grad school, pursuing an MD–PhD, professional school, or research during a medical residency, an undergrad research experience will help you prepare for those future research positions in part by acquiring basic (and possibly advanced) research skills, and by learning how to be a supportive and productive member of a professional research group. Furthermore, admissions committees for grad, med, and professional school programs notice when applicants have participated in inquiry-based activities and have consistently demonstrated critical thinking skills, intellectual curiosity, creativity, initiative, and the ability to produce results to their references. An in-depth research experience will provide numerous opportunities for you to develop and demonstrate these qualities to potential references. And if you're headed to grad school in the sciences, you'll likely gain an additional competitive advantage because many admissions committees in diverse fields and disciplines believe that participation in undergrad research is a reliable indicator of future success in grad school—much more so than an entrance exam.

But if instead your career goals include entering the research job market or a postbaccalaureate program directly after your undergrad degree is awarded, obviously acquiring research skills and having previous experience working with a research group will be advantageous.

Even if you ultimately choose a nonresearch career path, the interpersonal and professional skill sets you develop during your research experience can give a competitive boost during a job hunt. For instance, during your research experience, you'll likely have opportunities to demonstrate that you are reliable, work well with others, and are able to contribute to team and project goals, which will be reflected in your recommendation letters. Virtually every employer wants to hire the candidate with a proven track record in those areas. And if your long-term goals include an entrepreneurial track such as owning your own clinic or managing your own business, those skills will be essential for your success.

In essence, participating in a meaningful research experience can foster interpersonal development, give you an academic edge, create connections, and potentially lead to a fellowship or internship. Regardless of the career path you choose, participating in a meaningful undergrad research experience will provide opportunities to gain skills that you can use as a launchpad for it.

PROFESSIONAL DEVELOPMENT OPPORTUNITIES

Much of professional development can be classified into one of two categories: (1) accomplishments you can list on a CV (curriculum vitae) or resume, or (2) transferable skills that will enable you to succeed in future endeavors. And although items from either category might be included in reference or recommendation letters, transferable skill sets have the additional bonus of contributing to your success outside the research experience, in college, and in your chosen career path regardless of what it turns out to be.

CATEGORY 1—ACCOMPLISHMENTS THAT CAN BE LISTED ON A CV OR RESUME

In this section, we cover examples of research experiences with varying levels of time commitment and the *potential* professional development that can be gained from participating. **The benefits you'll *actually* gain will be connected to a variety of factors such as your research field,**

the training opportunities offered, your goals, and the time and personal investment you make in your research experience. We also want to mention that not all research groups offer experiences in all the summaries covered. One mentor, for example, might design a project that requires a time commitment of twelve hours per week whereas another mentor's project might entail only six hours per week.

In addition, it's reasonable to presume that the more time you spend conducting research (in hours and semesters), the more interpersonal and professional development opportunities you'll achieve. But the reality is more complicated. **How you use your time while you're in the research opportunity is also relevant as there is a distinct difference between *being* physically present in a lab fifteen hours a week and *working* in a lab for fifteen hours per week.** A student who spends a few minutes setting up an incubation or starting a rendering, for example, and then uses the next three hours to socialize on social media or stream shows might be in the lab, but that doesn't equate to participating in research the entire time. Or a student who spends most of their time washing glassware or observing labmates won't have the opportunities to gain many research skills or other benefits associated with working on a research project.

But consider a student who starts a procedure then uses the next three hours to learn more about how the procedure is relevant to their project or to actively assist a labmate with a technique. Overall, this student acquires more experiential learning than the previously mentioned ones. And although most undergrads will spend some time completing research-related chores, including setting up before a procedure and cleaning up after, if you're not doing much research as part of your "research experience," you're not going to maximize the benefits of being part of a research group.

In the examples presented next, certain achievements—such as expertise in a procedure or operating a piece of equipment—could technically be gained through most research experiences. We classified the accomplishments based on our experiences as mentors and on the opinions from colleagues in a variety of research fields when we asked, "When you write recommendation letters, what are you generally able to include for undergrad researchers based on the approximate time in hours and semesters the student was a member of the research group?" Therefore, consider this section a guideline, as your actual experience may vary, and remember that choosing a position with the greatest time commitment isn't enough to guarantee that you'll gain specific skills or benefits listed.

As you read through each category described next, highlight items you'd ideally like to be able to include on your resume or CV at the end of your research experience. You'll use this information later to identify, and eventually interview for, research positions that will help you accomplish your academic and personal goals. In addition, after joining a research experience, you'll be expected to be actively engaged in the processes of setting and pursuing your goals. If you wish to receive feedback from your mentor on how to best accomplish your aspirations (and we hope that you will), you can establish a foundation now by thoughtfully considering these next sections.

Short-Term Research Experience

This research experience involves the following:

- Three to six hours per week
- One semester or less time
- Working with the same research group

If you participate in a short-term research experience for a few hours a week and have minimal responsibilities, you might secure a letter of reference and be able to list some of the following accomplishments on your CV or resume:

- Familiarity with some research techniques or methods or expertise in a few
- Specialized knowledge
- Attendance at an undergrad research conference or symposium, such as one sponsored by a student group, department, or other campus program
- Presentation of your research project (poster or a short talk) at an undergrad research symposium
- Participation in a research experience

Longer-Term Research Experiences with Varying Commitment Levels

Note: We define a *full-time* research experience as conducting research for approximately forty hours per week for eight to ten weeks. These research experiences might be part of a fellowship or paid internship program. (We cover such opportunities in more detail in chapter 3, in the section "Summer Undergrad Research Programs (SURPs).") Such research experiences may involve the following:

- Ten to fifteen-plus hours per week during the semester
- Two to six semesters—ideally continuous and with the same research group or conducting *similar* research with multiple groups
- Might include a full-time summer research experience—possibly with the

same research group as your regular semester lab experience but can be with another group

OR

- One, two, or three full-time summer research experiences (without being involved in research during the school year). When only one experience is done, it's ideally started and completed with the same research group. When multiple experiences are completed, whether conducting research with a single group or several groups is best determined by the undergrad's goals.

OR

- A full-time research experience for a semester or a year that allows a student to focus exclusively on research without taking classes (or taking classes that are directed related to the research experience). These are sometimes classified as an internship or done through a study-abroad program.

An increase in time spent conducting research with the determination to maximize that commitment will likely boost a letter of reference to a strong, well-rounded letter of recommendation and possibly an authorship on a peer-reviewed journal publication. Additional accomplishments may include the following:

- Technical expertise and skills acquired in lab or field techniques, including troubleshooting skills
- Substantial experience using scientific software packages such as those used for clone design, DNA sequence manipulation, phylogenetic analysis, and statistical analysis, or others
- In-depth responsibilities associated with managing a research project
- Technical expertise with equipment
- Leadership skills from assisting in the training of other undergrads or other labmates
- Additional advanced responsibilities within the research group as guided by the lab culture
- Participation in an undergrad research conference, meeting, or symposium such as those sponsored by student groups, departments, or other campus programs
- Participation or presentation at a national or international undergrad research conference or a research conference where scientists at all levels present their projects

- Enough research completed to write a senior thesis or a paper for an undergrad research journal
- Scholarships, or awards for research, travel, or nonscience accomplishments (further details on these are covered later in this chapter)
- Being eligible for leadership positions in campus groups connected to research, such as student chapters of scientific organizations, or participation in outreach activities such as becoming an ambassador for undergrad research on your campus through a department or college

CATEGORY 2—SKILLS THAT HELP YOU PAY (FUTURE) BILLS

In this section, we discuss the transferable, professional, and interpersonal skills that you may develop, refine, and demonstrate during your undergrad research experience. These skills will help you be successful in the lab but will also contribute to your success elsewhere in college and during your career—even if you don't pursue one that involves conducting research. Which ones you'll gain will depend on a variety of factors, including your commitment level to the research and the opportunities available through your specific research experience.

Upholding a Commitment

Although we don't recommend sticking with a research experience that makes you miserable or risks your well-being, **involvement in an undergrad research experience will unequivocally show that you can stick with something** even through the rough spots (which all research projects inevitably have). These rough spots—which can range from struggling with a particular technique to pushing yourself though a boring phase or having to redo a procedure over and over and over again—often require mountains of patience. Whenever we poll undergrads on social media about the things they learned from participating in a research experience, the patience to stick with a project in difficult times is always a popular answer.

Working Well with Others

When you're working with peers, labmates, or your research mentor, getting along with others is essential, and the ability to do this well is critical to your success outside of college. **Those who have demonstrated the ability to coexist with all group members, respect the opinions of others, and engage in collaborative efforts are highly regarded by admission committees and employers.** And simply being a member of a research group doesn't automatically lead to the ability to work as a unified member of that team. For some students learning to navigate a diverse college landscape, getting along with others can be harder to do than expected. Sometimes this is because of outright prejudice or implicit bias. Misunderstandings or inappropriate behavior stemming from an intolerance of or insensitivity toward a coworker because of their culture, ethnic identity, sex, race, gender, gender identity, disability status, age, socioeconomic background, religion, or other reason wreaks havoc on morale and productivity within a group and amplifies a work culture of hostility and inequity.

Producing Results

You know that effort is important for success; however, the quality of the results you produce throughout your research experience are also important. Your mentor will devote time and lose some productivity to train you in research methods, safety procedures, and other research-related tasks. During this partnership, you'll have many opportunities to either make the effort to produce high-quality results or rush through a process just to get it done. If you consistently choose the former, you'll contribute results others rely on and develop the patience and self-discipline to help you succeed in your career.

Embracing Constructive Feedback

Everyone wants to work with someone who wants to learn. An undergrad who responds to their research mentor's constructive feedback with the determination to improve their performance will receive more help, more development and research opportunities, and experience more success than an undergrad who responds with defensiveness or outright ignores their mentor's guidance or instructions. **Use your research experience to strengthen your ability to absorb helpful feedback and apply it to improve your performance or approach to completing a task.** This practice will better prepare you for the future when critiques from a supervisor, and your reaction to it, will have the power to affect your career in a positive or negative manner.

Learning Research Skills

Everyone acquires new knowledge in college, but not everyone learns new skills. Through continued involvement in, and success with, your research project, you'll acquire a specialized skill set related to your research project. Even if you don't use all those same skills in the next phase of your career, every admissions committee and employer values candidates who have both the desire and the proven ability to learn new skills.

Demonstrating Leadership

If you're asked to help train or supervise other undergrads, the line on your CV might be "trained *X* number of people in technique *Y*," but admissions committees and potential employers know that it takes more than knowledge or proficiency to be an effective instructor—it requires leadership. **Having the responsibility of teaching a technique is good. Using that opportunity to refine and demonstrate your leadership skills is even better.** Outside your research experience, you might also choose to participate in peer-to-peer advising as part of an official program with the Office of Undergraduate Research or your major department. The more practice you have with these types of opportunities in college, the easier it will be to accept leadership opportunities after you graduate and start your career.

Following Instructions

It seems easy, right? However, many new undergrads try to "improve" a protocol by changing or skipping a step or sometimes make a change because they aren't paying attention. Changing an optimized research protocol usually results in a failed procedure and frustration for both the student and the mentor. **Undergrads who excel at following instructions and following standardized research protocols make more progress overall and spend less time redoing procedures that fail as a result of operator error.** These students are also more likely to be given advanced research responsibilities or the option of pursuing an independent research project. In every career path or professional position, you'll need to follow instructions. If you polish this ability during your research experience, you'll have the skills to go further in any career you choose.

Focusing in a Chaotic Environment

Research environments can be loud or chaotic, and focusing on the simplest task might be difficult—especially when your labmates' conversations are more interesting than a chore you're grinding through. **Learning to block out distractions is an essential skill to cultivate, and it takes a lot of practice to do it well.** We're not suggesting that you'll need to be perfect in this regard to be successful in undergrad research but that you might be able to use time spent in the lab to test strategies that might work for you. In many professions, you won't have a distraction-free environment to perform complicated tasks that are important to you (or your supervisor), and you won't have the option of putting on headphones to help you focus. Use your research experience to refine your concentration skills in a chaotic environment, and you'll be better off in all professional endeavors.

Honing Problem-Solving Skills

There are numerous ways to take ownership of your research project. Depending on your research experience, some ways might be by optimizing a new technique, troubleshooting a failed procedure, interpreting results on your own or with coaching from your mentor, planning the next phase of a study, or performing data analysis. **By thinking about the research on several levels, you'll develop the ability to think critically, evaluate problems, and seek solutions.** When a technique doesn't work, you'll learn to consider what likely went wrong and use your knowledge to determine what to do about it. In some cases, you'll learn to narrow the possibilities, determine which option is the best approach in the moment, and devise a strategy to move forward. You'll also learn to evaluate whether your strategy was successful or a result leads you to a different conclusion and the need to start again.

Developing Organizational Strategies

The better organized you are, the more successful you'll be—or the less you'll be forced to deal with the stress that accompanies being perpetually disorganized. In undergrad research, there will be numerous opportunities to establish, refine, and demonstrate your organizational skills. No matter the career path you choose, success will come faster if you build solid organizational strategies in college. Use your research experience to create the foundation that will help you succeed.

Planning

Whether you take a complicated protocol and break it down into small, manageable steps, or anticipate your future needs for completing fieldwork, **undergrad research will present myriad opportunities to refine your ability to plan.** There is no way to overestimate the advantage of being able to imagine yourself in a future situation and devise a plan to cover the most likely contingencies.

Embracing a Growth Mind-Set

In her book *Mindset: The New Psychology of Success*, psychologist Carol Dweck discusses that people who believe they can develop their skill sets to be successful have a growth mind-set. In all things research—learning new techniques, thinking critically about a problem, writing about research, and more—a growth mind-set is absolutely the key to success. You're not born with the skill sets or knowledge of how to do research, but you can develop these traits if that is what you want to do. It won't always be easy—often the process will be challenging and frustrating—but if you're determined to develop your research skills, you'll do it. In addition, when—not if—your research project hits a roadblock, **applying a growth mind-set will push you toward success.** There is no job or career where a growth mind-set is a disadvantage.

Developing Outstanding Communication Skills

Whether through question-and-answer sessions with your research mentor, or presenting a research poster on your project, or explaining your latest result at a group meeting, the sooner you start a research experience, the more opportunities you'll have to refine your communication skills. Even if you're one of the lucky few who feel comfortable presenting to large groups of strangers, you'll still need your audience understand your message. **Because there is no shortcut to developing outstanding communication skills, you should use your undergrad research experience to help learn to do the following:**

TAILOR YOUR MESSAGE TO YOUR AUDIENCE

How you'll discuss your research project will vary depending on whom you are connecting with and their background knowledge. For example, your research mentor will have advanced knowledge about your project, so you'll use technical language and known abbreviations when you summarize a result or discuss a strategy. However, if you discuss your research with a classmate who doesn't share your academic major, you'll use non-technical language so the conversation is meaningful to them. **The ability to adapt your message to your audience, without being condescending, is an essential skill, no matter what career path you choose.** It's a critical skill to polish especially if you choose a profession wherein you regularly interact with those who have less knowledge about a subject than you do, such as a professor, doctor, science journalist, or educator.

COMMUNICATE WELL UNDER PRESSURE

At the start of your research experience, even when they are patient and supportive, conversations with your research mentor may be nerve-racking. If you're asked to interpret a result, explain why a procedure failed, or give your opinion on something, you might initially panic and struggle to find the right words to form an answer. **Learning to steady yourself**

and give a clear and concise answer will initially be a challenge, but with enough practice, it will become easy. Really. You want as much experience with this as possible before you're an intern being grilled by the attending physician, an assistant professor teaching your first class, or an employee explaining to your boss why you deserve the promotion over a colleague who has been at the company longer. Take full advantage of the opportunities during your research experience to polish this skill.

REFINE YOUR NOTE-TAKING SKILLS

The ability to process information and re-create it accurately in your own words is a skill that you'll need in every profession, but one that only comes with practice. Currently, if your core note-taking technique is to supplement your professors' lecture slides, you could graduate from college without refining this skill. In a research experience, you'll need to take notes while your research mentor explains the upcoming steps in a procedure, the project objectives, a change in the protocol midway through, or other details—without the benefit of preprinted slides. In addition, when you record something in your notebook or take notes in your field journal, create a poster, or write a research report, you'll need to include enough details to be useful or make an impact on your intended audience but not so many that your message gets lost. Even if you're legally entitled to having accommodations in place to make your training equitable with these processes, the longer you're in a research experience, the more you'll practice these communication skills.

PREPARE FOR THE INTERVIEW

To take full advantage of your research experience, you'll want to present your research project every chance you get. Each time you present your research poster at a symposium or conference or discuss your research progress at a group meeting with your labmates, it will seem like an interview (perhaps *interrogation* is the more accurate word). Even if you do most of the talking and they ask only a few questions, these "practice interviews" will help you prepare for those oh-so-important interviews near the end of your college career. If you present your research frequently, by the time you go to your mock interview for med school, grad school, professional school, or the job market, your interview style should need only a little polishing.

Polishing Your Professional Skill Sets

Everything discussed previously is relevant to this point. Every opportunity that enables you to develop, refine, polish, or demonstrate your professional growth, also referred to as *transferable skill sets*, is an advantage. For many undergrads, a research experience is their first professional work environment and an important opportunity to navigate the expectations of a supervisor while thriving in a work culture. But whether it's an entirely new experience or you begin with a work history, an undergrad research experience is the perfect opportunity to gain skills you'll use to be successful in the next stage of your career.

Getting a Job after College

Depending on the skills you acquire during your undergrad research experience, **you may be employable as a research scientist after graduating with a bachelor of science degree.** Your options might include employment at a university or college research lab; working for a company, at a hospital, or in a government lab; or entering a postbac program as a paid researcher. Several of our former undergrads have been employed in professional research positions after graduating with an undergrad degree—some during a gap or personal year, others as a launchpad in their research career or before attending med school. Also, there are numerous nonresearch employment opportunities you might choose to pursue after undergrad.

INTERPERSONAL DEVELOPMENT OPPORTUNITIES

Some interpersonal and professional development characteristics overlap. For example, self-motivation is an advantage in every professional pursuit; however, it's equally important to achieve personal goals and to prioritize the activities that matter to you the most. It's also important to note that many interpersonal development characteristics support the development of others. For instance, being self-motivated makes it easier to stay self-disciplined, and practicing self-reliance makes it much easier to trust your instincts and make your own decisions.

Interpersonal development happens over time through new experiences and challenges within those experiences. As you develop your intra- and interpersonal skills, you'll learn to rely on your intellect, abilities, and strengths to solve problems and manage your life.

Participation in a research experience doesn't automatically guarantee interpersonal development, but it does give you continuous opportunities to challenge yourself. It's up to you to decide how to use those opportunities.

PERSONAL FULFILLMENT

Your research experience should be a meaningful and rewarding use of your time. Although there are several ways this can be accomplished, following are some examples.

Investigating Curiosity

Perhaps you liked doing the methods in a lab course and you want more exposure to those methods in another setting. Or you enjoyed a lab course but want to learn techniques in the associated field that weren't part of that experience. For some students, their desire to participate in research is ignited because their exposure to science made them curious about other opportunities.

Making a Difference or Contributing toward a Common Goal

For many undergrads, their research experience is an opportunity to find personal fulfillment by participating in something bigger than themselves. No matter the science, if you're studying population genomics of coral reef invertebrates, researching a disease, or trying to understand an important pathway, **knowing that your contributions could have an impact beyond your personal gain is unequaled.**

Connecting with Your STEMM Identity

Your personal STEMM identity is the perception you have of yourself in relation to science, technology, engineering, math, and medicine. For most students, their STEMM identity is multifaceted and complicated, but it's influenced by answers to questions such as the following:

- Who am I as student or researcher in science?
- Where do I belong in science?
- What are my capabilities, skill sets, or ways I use technology?
- What goals am I interested in pursuing through science?
- How do I think others perceive me as a person in science?

Right now, maybe you're unsure where you belong in science, possibly because you're a student with an underserved or marginalized identity or don't know any scientist role models who share your social identities. Or maybe you have an idea of how you would like to contribute to science through a particular profession but aren't sure whether those goals are achievable. **Conducting undergrad research is a process of self-discovery.** Through skill development (not just the technical skills but also those requiring critical thinking, and creativity, for example), virtual or in-person interactions with mentors and role models in the sciences, and other opportunities as described in this chapter, you will become better able to create, understand, and evolve your STEMM identity.

Defining or Clarifying a Career Path

Is a career in research right for you? Are you trying to decide among med school, grad school, and another professional school? Are you pursuing a science degree but aren't sure what to do with it after graduation? Are you in the right discipline? Have you declared the right major? Maybe research on a microbiology project will inspire you in a way that your major never has. Perhaps after spending time in a research experience just to try it out, you'll consider pursuing a grad degree or MD-PhD or realize that you'd like to consider a nonresearch career in science policy, health communication, or civic ecology. A research experience can help you clarify your career path.

There is another potential advantage that accompanies many research experiences: Engaging in authentic research provides exposure to the daily grind that is part of most projects. Although this might not seem like a benefit, the challenges of navigating repetitive and time-consuming procedures, managing failure in research, and consistently needing to improve technical skill sets can be enlightening. **Participating in undergrad research can help you determine whether you're on the right career path or if you should consider taking a detour.** Even if you don't discover something groundbreaking through your research project, you'll likely discover whether you might be interested in a career in science—even if it's not in research.

Boosting Self-Confidence

In every area of interpersonal and professional development, there is a potentially hidden bonus—the increase in self-confidence that comes from earned accomplishments. *Your* accomplishments. As you gain technical skills, learn how to explain the details of your project to your family and peers, and solve complex problems related to your project, you can derive a great deal of satisfaction from knowing that you've accomplished something that wasn't easy.

Understanding Self-Motivation

Self-motivation is your reason or motive for participating in anything—a road trip with friends, a study group, an extracurricular activity, or a research experience. **Understanding your self-motivation is an essential step in taking control of your life and your future.** The key to self-motivation is making sure your reasons are truly about your desires, your wishes, and your goals. It's much more difficult to achieve goals if you don't have a clear understanding of why you should pursue them, or if they are actually someone else's goals. Once you learn to evaluate (and occasionally reevaluate) your self-motivation for undergrad research, it will help you apply the same accountability to all your activities.

BUILDING SELF-MANAGEMENT SKILLS

Self-management is the ability to be persistent in planning your life and to follow through with activities that are important to you—especially when faced with challenges and frustration. **Developing self-management skills will help you identify and pursue your passion and build the life you want.** The following are some of the self-management skills you can gain from an undergrad research experience.

Strengthening Self-Discipline

Self-discipline is the ability to pursue your goals even when it's difficult, frustrating, or boring, or when it would be easier to just give up. When you hit a rough spot with research, and your self-motivation isn't enough to get you through it, self-discipline will help you get the work done regardless of how uninspired you feel. Self-discipline is essentially adopting the philosophy, "I'm going to keep working toward my goals and stay focused, even though it's hard." It's the ability to find the resolve to keep pushing forward even when you don't feel like it in the moment—not because you'll "get in trouble" if you don't, but because you have the perseverance to accomplish your goals. **If you use your research experience to strengthen self-discipline while in college, you'll be less likely to give up on your dreams when your life hits a rough spot.**

Improving Self-Reliance

Self-reliance is the ability to use your personal resourcefulness to complete your responsibilities and accomplish tasks. In a research position in a *wet lab* (a laboratory equipped with liquid chemicals and reagents), for example, this might cover everything from learning where reagents are stored to knowing what do to when you arrive at the lab each day, and ultimately to designing your own experiments, if that option is open to you. **Self-reliance involves making a conscious decision to try to solve a problem or answer a question yourself instead of instinctively asking someone else for help.** But it's important to understand that you can be self-reliant and still receive assistance from others because the two aren't mutually exclusive. No one will expect you to know everything, and in some cases it is indeed wiser to ask someone with experience than to make a guess when working with something that could literally blow up in your face. The key is to balance what you can do for yourself with requests for help from others. It takes time and practice to be able to do this well. A research experience can guide you toward self-reliance, but it will be up to you to embrace the challenge to improve it. The more you practice self-reliance, the easier it will become and the more it will be a part of your approach to life.

Developing Strategies to Cope with Failure and Disappointment

The one guarantee about research is that sometimes things fail. In a wet lab, it could be a technique, an experiment, or a straightforward should-have-been-easy-to-do dilution. In a *dry lab* (one pursuing work typically done at a desk or table and that doesn't involve liquids), a power failure might stop a computer-generated model from completing after several hours, requiring you to redo the work. And field researchers often have unexpected complications—such as wildlife or inclement weather that destroys their site midstudy. Our point is that **occasional frustration and disappointment represent the reality for most researchers, so developing coping strategies is essential to be successful.** If you're able to do this consistently throughout your research experience, you'll develop and reinforce positive strategies to use in your life outside college as well.

TAKING RISKS TO OVERCOME FEAR

Because most new experiences involve getting out of one's comfort zone, most undergrads experience a "healthy fear" at the start of a research experience. This fear, when properly managed, can lead to interpersonal development and lessons learned about oneself along the way. **The most common fears undergrads overcome through involvement in an in-depth research experience are the following:**

The Fear of Starting Something New

Especially at the beginning of a research experience, the lab is an intimidating place—even when everyone welcomes you to the team. At the start, you might be afraid that everyone is judging you on a personal level (they aren't). You might be afraid of interpreting your results for your research mentor (it will get easier with practice). You might be afraid that you'll never become self-reliant like the other undergrads are (with hard work and a willingness to learn, you will). **Through patience and determination, you'll learn to push through your initial nervousness and begin**

to feel like a member of the research group. Then the next time you join a club, start a volunteer experience, or begin an extracurricular activity, it will be easier because you'll already have experience transitioning to a new environment at the college level.

The Fear That You Don't Belong in Science

You might or might not be familiar with the term *impostor syndrome* (which is sometimes called *impostor phenomenon*), but chances are you're acutely aware with the associated feelings. The concept, introduced in 1978 by psychologists Pauline R. Clance and Suzanne A. Imes, describes internal feelings of inadequacy even when it's evident that someone is competent or even an expert. Experiencing impostor syndrome causes you to amplify internal thoughts that you're not smart, talented, or skillful enough to be successful in your academic program (or elsewhere) or the tendency to downplay your accomplishments as luck. These self-sabotaging feelings of inadequacy can be so powerful that they negativity impact your STEMM identity.

Impostor syndrome affects nearly everyone but can be harder to navigate for a student who thinks they are the only one who experiences it, particularly for an undergrad from a marginalized or underrepresented community in STEMM; for a student who is undervalued, harassed, or discriminated against in either subtle or obvious ways within their academic community; and for a student who doesn't personally know any scientists who share their culture, ethnicity, gender identity, socioeconomic background, or other identities. **Conducting undergrad research won't cure impostor syndrome,** in part because for most people moments of self-doubt resurface periodically throughout their career, **but through your involvement in a meaningful undergrad research experience, you can gain technical and interpersonal skills that will help you realize that a career in science is open to you, if that is what you want.** We also hope that your labmates and mentors will offer helpful suggestions on tackling impostor syndrome.

The Fear of Making the Wrong Research Decision

Determining your options when designing an experiment, choosing which procedure to use, or troubleshooting a failed technique is only the start of accomplishing your research objectives. Making the decision of which option to choose can be much harder. By understanding your project and relying on your knowledge, skills, and abilities, you'll learn to consider the "what ifs" to create strategies, make decisions, and commit to an action plan. **Overcoming the fear of making the wrong decision won't be easy, and it won't happen all at once, but through a commitment to your research experience, it will happen.** When it does, you just might find that the confidence to make decisions sticks with you outside the lab as well.

The Fear of Making a Mistake

Overcoming the fear of decision making and the fear of making a mistake go together. For most new research students, after deciding which research strategy is the best to pursue, the greatest source of stress emanates from the fear of making a mistake. We assure you; it will happen. However, when it does, **you'll learn firsthand that making a mistake can't destroy you, and it doesn't mean that you're unintelligent or incapable of getting something right the next time.** If you learn to put research mistakes in perspective (it's not the end of the world), and analyze the situation (why did it happen, what can you do to prevent a similar one in the future), it will help you realize that letting the fear of making a mistake prevent you from taking action can be the bigger problem.

The Fear of Seeming Unintelligent

At the start of a research experience, it might be difficult to ask for help with a protocol or to interpret a result for your research mentor. Especially at the start, you might feel that you're supposed to know more than you actually do and worry that your research mentor will be upset if you ask too many questions. (This is a common fear among new undergrad researchers.) But if you're working with a supportive mentor, **each time you ask a question about a procedure, or answer a query about your research results, or bring a mistake to your mentor's attention, you'll make progress on controlling that fear.** The more often you put yourself "out there" during your research experience, the more your self-confidence will grow, and the less foolish you'll feel. As a bonus, this growing confidence should also make it easier to interact with professors in office hours (if that currently makes you nervous) and to ask questions during a lecture or lab class.

The Fear of Change

Especially if wet lab benchwork is part of your research experience, at times it will seem as if your project is an endless string of spontaneous events that you can't control: Your cultures didn't grow. Your plasmid didn't cut. Your cell lines got contaminated. A new result changes the direction of your project. Your research mentor seemingly randomly instructs you to process samples with a protocol you've never done before, and they won't be around to help or answer your questions while you do it. It can be disconcerting not to have a defined plan of what will happen each day in lab or to need to change research strategies in the middle of an experiment. **If you can learn to embrace change in research, instead of being frustrated or overwhelmed, it might help you develop strong personal adaptability.** This will make you a happier person overall in the lab, out of the lab, and in life.

DEVELOPING AN ACADEMIC AND LIFE BALANCE

To get the most out of your college experience, you'll need to develop an appropriate academic and life balance. In this section we present several ways your undergrad research experience can help you achieve this commonly sought-after goal. If you're working your way through college, you're already aware of how important time management is, so this section may be more of a reminder than new information.

Using Organizational Strategies Learned in the Lab as a Model for Personal Success

It's not easy to get and stay organized—that's why there is an entire industry dedicated to tips and tricks on how to organize one's life and career. For many undergrads, organizational skills develop slowly, through a process of trial and error, often punctuated by stressful moments caused by overcommitment. Although you won't do protein analysis or complete an actin polymerization assay in your life outside of your research project, **if you choose to apply some of the same organizational strategies that help you get research done to your life pursuits, you'll get further ahead with less stress** than someone who doesn't have the advantage of actively participating in a research experience.

Refining Time Management Skills

Having enough time to do the things you want to do, as well as the things you're obligated to do, doesn't happen by accident—it happens through planning and effective time management in all areas of your life. **Depending on the parameters of your research project, where it takes place, and the hours required, through necessity you'll develop and refine your time management skills.** As a result, you'll spend less time cramming for exams and finishing assignments at the last minute and more time enjoying your overall college experience.

Learning to Prioritize the Activities That Are Important

With only twenty-four hours in a day, every activity you choose comes at the cost of doing something else. Every class you take, study group you join, YouTube video you stream, or volunteer opportunity you pursue requires you to take the time from somewhere. **When estimating your availability for a research experience, you'll need to consider each activity and decide whether what you'd expect to get out of it would be worth the time you'd spend on it.** On the surface, this may seem a simple task, but for many students it can be incredibly difficult when faced with opinions (or pressure) from a well-intentioned friend, family member, or classmate. *Learning to prioritize your time also means learning to trust your own judgment about how you should spend it.* This is such an important activity to learn to do that we've dedicated a full section to the process in chapter 4.

INTELLECTUAL PURSUITS

Research has intellectual challenges beyond learning facts, understanding theories, testing hypotheses, and reading current scientific literature. Pushing yourself to connect with your project on an intellectual level beyond the absolute minimum will lead to some of the following personal benefits.

Driving Intellectual Curiosity

In most research experiences, you can simply choose to be a passive participant or you can make an intellectual investment in your project. In other words, you could choose to approach your research project as a series of tasks to be carried out and checklists to be completed. Alternatively, you could choose to understand why a particular technique or approach is relevant to your research, why the result of a procedure is important, or why your project is relevant to the citizens of a specific geographical area or the world. **The choice is up to you, but doctors, researchers, and entrepreneurs who are interested in the hows and whys are also the ones who discover or invent creative solutions to problems.** Use your research experience to nurture your intellectual curiosity (and creativity) early, so it's second nature by the time you need to rely on it in your life and career outside college.

Fostering Creativity

Creativity is not limited to the arts. The experiential learning that accompanies an in-depth research experience will provide many opportunities to develop creative-thinking skills—such as the ability to shift your perspective to consider new, possibly unconventional approaches to solve a problem or theorize an alternate yet equally plausible explanation for an unexpected research result. **Those who develop creative-thinking skills are less likely to stagnate in a professional position, more likely to find personal fulfillment and significance in their life**, and more likely to introduce inventive solutions as needed in both. And, yes, unlocking your creative problem-solving skills will absolutely transfer to your life outside research and well beyond undergrad work.

Sharpening Critical Thinking

Research teaches you how to evaluate the outcome of experiments and procedures and formulate possible explanations. To be successful, it's essential to remove your personal biases from consideration, analyze the relevant facts, and set aside the inconsequential ones. Doing so allows you to make conclusions and decisions based on the evidence and facts—not on what you want to be true. An in-depth research experience also trains you to notice when there isn't enough evidence to support a hypothesis or give weight to a conclusion. As you learn to do this through your research project, the ability will inevitably extend to your personal life. **You'll become a more discerning citizen and gradually begin to expect higher standards of evidence from your friends, coworkers, reporters, politicians, and internet bloggers before accepting a statement as a fact.** Essentially, you'll be less likely to take someone's word for it and be more likely to recognize when a conclusion isn't supported by the evidence (or lack thereof).

ACADEMIC ADVANTAGES

Beyond professional and interpersonal development, you could potentially gain academic advantages by participating in undergrad research. These benefits will support you throughout college, and some will continue to pay off after you graduate. Some advantages include the following:

ENGAGING IN THE SCIENTIFIC PROCESS

As an undergrad involved in a research project, you're immediately involved in the vanguard of research, giving you unmatched exposure to the process of science—how discoveries are made and confirmed and how information is obtained and synthesized. Perhaps you'll be part of the joy of discovery and experience firsthand how new and old information are combined to construct new models of how processes work.

MAKING A CONTRIBUTION OR A DISCOVERY

While it may be unlikely that you'll discover something absolutely new, **the discoveries you do make during an undergrad research experience will likely help solve a problem or answer a question.** Whether you answer a specific question about how two genes interact or contribute information into the global effects of food webs, your contributions will add knowledge to the universe.

EXPERIENCING SCIENCE IN ACTION

A research project is the ultimate in experiential learning. Although you could learn about cell division from studying it in a lecture class, the key concepts will be unforgettable if your research project involves manipulating cell division and quantifying the phenotypic effects at the research bench. Aristotle said it best: "For the things we've to learn before we can do them, we learn by doing them."

EARNING COURSE CREDIT

The advantages of taking research for course credit vary substantially, depending on your college and department. In some departments, undergrad research factors into a grade point average (GPA), and in others, research for credit isn't offered. Some students register to ensure that research is listed on their transcript or to remain a full-time student while taking one fewer lecture class during the semester. Registering for research credit may also be required to meet academic requirements, such as pursuing an honors thesis or applying for certain fellowships, or might be required by the professor in whose lab the research is done. Whether or not you take research for academic credit is only partially up to you because your mentor, department, or college might have requirements one way or the other. We cover this topic more thoroughly in the chapter 3 section "Deciding Whether to Register for Course Credit for Research."

WRITING A THESIS OR AN UNDERGRAD RESEARCH JOURNAL PUBLICATION

Writing an undergrad thesis or publishing a paper in an undergrad research journal is an achievement. It will be included on your CV, and you'll have the personal satisfaction of funneling your hard work into a publishable manuscript. Beyond that, there may be other tangible rewards. In some academic programs, publishing an undergrad thesis or research paper paves the way for graduating with honors, earning a degree with distinction, or qualifying for certain scholarships or awards.

BOOSTING YOUR COURSEWORK

Your research experience can give you an advantage in classes and supplement your classroom knowledge. Depending on your research project, you might learn to use research tools, methods, or techniques that are covered in an advanced lecture or lab class. To have even a basic introduction before studying them in a class is an advantage. For example, if you use polyacrylamide gel electrophoresis during your research experience, learning about it in your biochemistry class will be easy.

PREPARING FOR THE MCAT OR GRE

Your research experience may include numerous opportunities for you to critically evaluate the outcome of a technique or analyze the data from an experiment. **The more you hone your critical thinking and analytical skills, the better prepared you'll be for tests that include these types of questions.** Several of our former undergrads have commented on how useful the skills learned during their research experience were on the reasoning and problem-solving sections of the Medical College Admissions Test (MCAT). As for the Graduate Record Examination (GRE), many departments and grad programs are dropping the exam as a requirement for grad school. However, it's possible that you'll decide to apply to a program that hasn't made this change and therefore will need a certain GRE score to be competitive for admission.

INTRODUCING YOURSELF TO A GRAD SCHOOL-LIKE EXPERIENCE

If you're considering grad school, conducting undergrad research can be immensely helpful in acquiring some tools and knowledge that will better prepare you for the rigors and challenges in a future program. Participating in research will introduce you to a professional lab environment, familiarize you with the inner workings of wet or dry labs or field research, and aid you in acquiring interpersonal and research skills. Although you shouldn't consider undergrad research an exact replica of what your experience in grad school will be like, it can be a valuable introduction.

EXPLORING SEMINARS OUTSIDE YOUR ACADEMIC BUBBLE

If you're a member of a research group, you'll be more likely to attend research seminars and symposia hosted on your campus. This is, in part, because your in-lab mentor or principal investigator (PI) might encourage attendance or because you decide there is value in learning more about some topics from experts. Imagine no quizzes or exams—simply learning because you're genuinely interested in the topic. Attending will make an impact on you whether a seminar is related to your current research or projected career path or is something that has personal relevance. The longer you participate in a research experience, the more inspiration you'll draw from attending such events.

CONNECTIONS—PROFESSIONAL, PERSONAL, AND MORE

In a research environment, every interaction is an opportunity to make a personal or professional connection. From technical assistance on a procedure or instruction on how to use a piece of equipment, to learning about events on campus, getting career advice, or building lasting friendships with others, the connections you make with your labmates can be a source of both knowledge and inspiration. Next are several possibilities that we hope will be part of your research experience.

PROFESSIONAL CONNECTIONS WITH OTHER UNDERGRADS

Some of your undergrad labmates may have career goals similar to yours. They might plan to attend grad, med, or professional school (perhaps after a gap or personal year), start in a professional research position directly after graduation, or use their STEMM degree in any number of professions that don't include research. These labmates might have advice and perspectives on application processes, classes, professors, campus social support groups, volunteer experiences, club memberships, and campus resources for well-being, food insecurity, and more. (And you might choose to share your experiences with labmates in turn.)

When you solicit advice and opinions from a labmate, it's often easier to evaluate that advice than it is from an anonymous person posting on the internet, partly because you've observed the work habits, academic commitment, and self-management of a labmate. If you know that a certain shadowing or volunteer experience helped them determine that pediatric oncology is now their true path, or that a particular professor in a particular class was too demanding and graded unfairly, you can use your personal observations to determine how much value to place in their advice and opinions.

Also, labmates who won't be competing directly with you for a future professional position (because they are farther ahead in the academic cycle or have chosen a different path) are sometimes more likely to share honest opinions about problems and pitfalls they faced or opportunities they seized during their early undergrad years. However, we wish to emphasize that most of your labmates will freely share information even if they believe you'll be direct competitors. But from our own experiences and stories shared with us by colleagues, we know that on occasion unfortunate competitive dynamics play out in a research environment.

PERSONAL CONNECTIONS WITH OTHER UNDERGRADS

On a personal level, you might form close friendships with other undergrad labmates. Although you'll want to do most socializing outside your research hours, many students find that the personal connections they make with other group members are substantial and can become lifelong. Even years after graduating, many of our undergrads have told us that they remain in contact with everyone (or almost everyone) who was in the group at the same time. Although most stay connected primarily through social media, several still reconnect in person for major life events.

PROFESSIONAL CONNECTIONS WITH GRAD STUDENTS, RESEARCH STAFF, AND THE RESEARCH PROFESSOR

Those who have "been there" can be a valuable source of advice because their perspectives reflecting on their undergrad experience are rooted in their successes and failures. Grad students early in their graduate degree program aren't too far from their undergrad days to remember the mistakes they made, the successes they had, and what they would definitely do again (or never again). Professional researchers tend to focus on career building and know what opportunities helped them get ahead and which ones they wish they would have invested more time or effort in. The professor who directs the research program you're a member of was an undergrad at some point and has experience navigating their career and mentoring other researchers. **These labmates have insider information that could help you excel at research and during your college career.** Plus, it can be immensely helpful to ask those who are farther along in their career path about why they chose it, what they wish they would have done differently, or what they are glad that they did—even if you're not interested in the same route.

However, even if the research team you join has a mixture of students and professionals at multiple career and life stages, the advice from a specific individual may be something that worked for them but is not tailored to the reality of your experience. For example, a labmate who has experienced more privilege-based advantages because of their race, sex, gender, ethnicity, or socioeconomic status might be unaware of how some strategies that worked for them might not be as easy for you.

CONNECTIONS WITH YOUR IN-LAB RESEARCH MENTOR

Although we hope that you develop productive mentoring relationships with several labmates, the connection you establish with your in-lab mentor might one of the most significant. In addition to helping you work through challenges with your project, they might advise you of opportunities outside of your research experience that could be beneficial to your professional or interpersonal development. For some undergrads, their research mentor becomes an important part of their overall success in college and remains in their life after graduation.

CONNECTIONS WITH OTHER PROFESSORS

When I (DGO) was an undergrad, a defining moment came during a conversation when a professor, who was not my research advisor, asked me about my research project. This was the first time I had a meaningful conversation with a professor about something other than lecture or course material. To get to discuss *my* research project was exciting, and although I knew it wasn't on the same level as one of their colleagues or a grad student, I still felt important and left the conversation feeling inspired. As a professor now, the interactions I've had with the undergrad researchers in the lab, and students I connect with at office hours, indicate that this is a universal, cross-generational feeling.

Participating in a research project—and understanding it enough to have a meaningful conversation with a professor (or anyone else, for that matter)—is a personal accomplishment. And when the conversation is authentic, it helps you connect with professor and might help set the stage for a recommendation letter even from a nonresearch professor.

CONNECTIONS THAT GROW

When grad students and professional researchers move on to their next career position, they may remain nearby or could relocate across the country or even across the world. Depending on the timing and the funds available, you might be able to visit a former labmate for an additional research experience in their new institution, or they might recommend you to one of their colleagues. (Sometimes as a former labmate's professional connections grow, your potential network does as well.) If you take a personal year (or several) and decide to work as a research scientist, or eventually determine that grad school or a long-term professional research position is your career path, reconnecting with a former labmate might lead to an interview, a job offer, or a referral to another research group. Even if the former labmate isn't responsible for the hiring decisions in their new group, they can serve as a reference if you ask them or give them permission to do so.

CONNECTIONS THAT INSPIRE

When you attend a conference or symposium, you'll be exposed to new ideas, concepts, and approaches in research. Some will be directly related to the research you do, and some will be tangential. **At any meeting, you're likely to learn more about how the research you do connects to an even bigger puzzle beyond your research group and to the world at large.** We hope that you already know that you're answering an important question, but even so, it's inspiring to connect with someone outside your group who is excited to learn about what you've discovered. We're not suggesting that it's always easy (or possible) for an undergrad researcher to pay for a conference's registration fee, travel expenses to the venue, possibly lodging, and other associated meeting costs. For those students, financial support in the form of travel funds from their home department or a travel stipend from a funding source managed by their research professor might be essential to facilitate attending an off-campus event. Regardless of the availably of these resources, most students also will benefit from applying for a travel award or reduced meetings fees from conference organizers. In particular, undergrads from any country can apply for travel awards at conferences sponsored by the Genetics Society of America (https://genetics-gsa.org). And at the time of this writing, the American Chemical Society (https://www.acs.org) was giving as much as a 40 percent discount on conference fees for current members.[1] But once you join a research experience, you'll connect with your mentor to discuss attending and paying for conferences.

CONNECTIONS THAT ENLIGHTEN

Some scientific conferences offer workshops for undergrads on career development, including acquiring funding for grad school, or opportunities

1. American Chemical Society (ACS), "ACS Membership: Choose Your Package; Premium," accessed June 3, 2022, https://www.acs.org/content/acs/en/membership.html.

to network and interview for jobs in industry. The workshops generally include information applicable to students at all academic stages and provide guidance for professional life after college. As a bonus, if you make connections with other conference attendees, you could learn about specific opportunities, internships, or a career path you've never considered (or even knew existed).

Although many conferences welcome undergrad researchers, some meetings place a specific emphasis on student enrollment and participation. The Annual Biomedical Research Conference for Minoritized Scientists (formerly the Annual Biomedical Research Conference for Minority Students), which is managed by the American Society for Microbiology, for example, includes scientists at multiple career levels, but as of the time of this writing, most attendees are undergrads or those conducting postbaccalaureate work.

CONNECTIONS THAT TAKE YOU PLACES

After you've started your research experience, you'll be more likely to consider participating in additional research and research-related activities. You might investigate national or international research opportunities that include a summer, semester, or year abroad. Perhaps, as graduation nears, you'll consider pursuing a master's degree or joining a postbaccalaureate program during a personal development year on your home campus or a place that you've never even visited! With research experience and a specialized skill set come the knowledge and confidence that will make it more likely for you to consider (and pursue) new adventures.

POTENTIAL FINANCIAL REWARDS

Unfortunately, financial rewards for conducting undergrad research aren't ubiquitous, but participating might open the door to some possibilities. In this section, we discuss some potential financial awards that you might be eligible for as an undergrad. However, you might also want to check with the campus Office of Undergraduate Research, Financial Aid Office, Honors Program Office, and the departmental office where you choose to conduct research. At some academic institutions, the Dean of Students Office also has guidance in this area. Eligibility for some awards, such as scholarships, might be reserved for students who are currently involved in research. Others, such as paid fellowship programs, might be exclusively for undergrads who haven't conducted research, are at academic institutions with limited research opportunities, or are members in one or more underrepresented or underserved communities. Although most of the items in this section would also go on your CV or resume, we've included them here because it allows for a longer description.

FELLOWSHIPS AND INTERNSHIPS

Paid research fellowships and internships are available at colleges, universities, national labs, government labs, centers, institutes, museums, nonprofit organizations, and other entities. **Many of these programs are open to undergrads regardless of their home institution, although most have strict eligibility requirements.** National and international opportunities exist for enthusiastic undergrads who submit professional application packets with solid recommendation letters, although they are highly competitive. These fellowships and internships can be the ideal opportunity for an intensive research experience over the course of a semester or a summer and often include the additional benefit of room and board and a tuition waiver. We cover more on applying for summer research programs in chapter 3.

SCIENCE-BASED AWARDS AND SCHOLARSHIPS

For most awards sponsored by a department on campus or through the Office of Undergraduate Research, there will be eligibility requirements such as current involvement in a research project or research field or subfield, but it's worth sending an email to inquire about the possibilities. (Also, check with your in-lab mentor because they may be aware of interdepartmental awards.) Scholarships can range from funding to purchase research supplies or defray the cost of attending a scientific meeting to a stipend to help cover living expenses while visiting a research group that collaborates with yours to learn a technique that you'll later use on your project. Also, some on-campus research symposia grant awards to undergrads who win poster presentations or short talk competitions that can be spent however the recipient decides—even if it's a nonresearch expenditure.

PAYCHECKS

Paid research positions, whether they take place at a field site or a lab and whether they are done during the regular semester or a summer term, are often more competitive and less common than volunteer or research-for-course-credit opportunities. Unfortunately, this inequitable system makes it more difficult (and sometimes impossible) for low-income students who are working through college to participate in a research program. Although some PIs pay undergrad researchers, outside a fellowship or scholarship program, securing a paid position is often more likely if a student has a financial aid award, such as a federal work-study award.

However, note that many academic institutions prohibit earning class credit and a *paycheck* at the same time. So, if you do secure a paid research position, ensure that you know how it will affect your future financial aid, if at all, and whether you'll owe taxes, so there are no unpleasant surprises.

TEACHING ASSISTANT PROGRAMS

When you have research experience, you'll be more a competitive candidate for summer opportunities to teach at university-sponsored science programs for middle or high school students. These programs pay undergrads to serve as instructors for several weeks during the summer. Programs range from teaching specialty topics, such as DNA-handling techniques or clinical pathology techniques, to full disciplines, such as genetics or cell biology. Although the programs vary, responsibilities typically include teaching lab techniques, a lecturing component, an advisory component, and a supervisory component. Most programs pay a stipend as well as room and board as financial compensation. Although sometimes eligibility requirements include being a junior or senior in a major connected to the subject matter, most positions are open to students from any academic institutions.

NONRESEARCH SCHOLARSHIPS AND AWARDS

The same professional and interpersonal strengths that you develop and refine in an undergrad research position are valued by selection committees for nonresearch awards. Participating in undergrad research also shows that you invest in your college experience and are involved on your campus, which is helpful on some scholarship, fellowship, and award applications. Essentially, include your involvement in research and associated accomplishments on every award you apply for, regardless of whether it is science related. In addition, check with the Financial Aid Office about potential onetime awards for which you might be eligible.

RECOMMENDATION AND REFERENCE LETTERS

During your undergrad career, and near the end of it if you choose to apply for academic programs or jobs, you'll need to obtain recommendation letters from professors who have some familiarity with your skills and accomplishments. You'll get stronger letters from them if you give them reason to remember who you are and what you've done.

FROM YOUR RESEARCH PROFESSOR

One of the tangible benefits you can earn from a research experience is recommendation letters. As a member of numerous faculty search committees, grad admissions, and fellowship committees, I (DGO) have read countless letters. **By far, the recommendation letters that make the largest impact include direct observations by the letter writer about the technical skills and possibly interpersonal strengths demonstrated by the candidate.**

Participating in research can provide the opportunities to repeatedly demonstrate interpersonal growth, professionalism, motivation, self-discipline, character, cultural competence, creative thinking—and so much more. Essentially, if you can learn it or refine it, you can demonstrate it. Almost everything listed in this chapter can be included in a supportive letter on your behalf. However, as with the other benefits of research, the recommendation letter you earn will depend on your level of commitment and the opportunities that accompany it. If your "research" opportunity does not include any actual research, then your mentor won't be able to state otherwise in a recommendation letter.

As an aside, it's appropriate to ask your research mentor for a recommendation letter when you apply for nonscience scholarships or awards.

FROM YOUR LECTURE PROFESSORS

Everything you do to enhance your professional development and gain the associated transferable skill sets through a research experience demonstrates your commitment to your education and future career path. When a professor agrees to write a recommendation letter on your behalf, any professional development or activities could be used to enhance their letter even if your primary contact with them was as a lecture student. Several colleagues have shared that they are generally able to write stronger recommendation letters for lecture students who participated in meaningful research experiences.

ONE NOTE ON PREPARING FOR LETTERS

After agreeing to write a recommendation letter, many professors will ask for a self-assessment (possibly as bullet points) detailing some research achievements or interpersonal strengths you demonstrated in your research experience to date. For many undergrads, this self-assessment is one of the most difficult "assignments" in college. To help with the process, review this chapter and highlight each interpersonal or transferable skill you gained or demonstrated for each reference. (Use a different highlighter color for each reference.) Then when you start to write the bullet points, choose the ones where you can include a specific story or detail on how you demonstrated it. But note, remember that you probably won't have demonstrated all the previously mentioned characteristics to every one of your references, and that is okay. You also only need to include three to five points for each letter writer—even if you demonstrated more.

2

An Introduction to STEMM Research, Research Groups, and Lab Cultures

WHAT IS SCIENTIFIC RESEARCH?

Scientific research is the systematic study of a subject to learn new facts and reach new conclusions. In the sciences, this includes investigations in many disciplines, such as mathematics, biology, physics, ecology, and chemistry. Research may also span disciplines and include theoretical, experimental, translational, computational, or other methodologies. And the methods used to conduct research in the sciences can involve conducting or collating surveys, doing fieldwork, performing techniques in a lab, creating computer models, conducting clinical studies, and more.

Generally, research is either hypothesis-driven or discovery-based. *Hypothesis-driven research* involves creating a hypothesis and then doing work (an experiment) that tests it. The results from the experiment will either support or disprove the hypothesis. *Discovery-based research* usually involves creating or analyzing large-scale data sets or resources that can be used by the wider scientific community to generate new hypotheses. Individual research groups often use both approaches.

This section covers a broad overview of research, and in some cases a generalization of some aspects of research that you or a classmate might be involved in as an undergrad researcher. We hope that by covering these topics, we will demystify the overall concept of undergrad research to our readers who have yet to begin a project and those who are new to one.

SOME INSTITUTIONS THAT CONDUCT RESEARCH

Scientific research is done at a broad range of academic institutions. Some examples are community colleges, research-intensive institutions, teaching-focused institutions, Minority-Serving Institutions (MSIs), Hispanic-Serving Institutions (HSIs), Tribal Colleges and Universities (TCUs), Historically Black Colleges and Universities (HBCUs), and Asian American and Native American Pacific Islander–Serving Institutions (AANAPISI). But it might surprise you to learn that research also is carried out at hospitals, at research institutes (often these are associated with a specific academic institution), at museums, at zoos, in industry, at non-profit organizations, and, in the United States, in national government labs. This is not a complete listing of sites where scientific research is conducted.

Some of our readers will conduct research during the calendar year while taking classes, and others will join research programs exclusively during the summer months. Undergrads who conduct research on their home campus during the semester or quarter might remain with the same research group for their entire experience (possibly years!) or they might work in several labs during that time. And undergrads who spend their summers away from their primary institution might participate in multiple research-intensive internship programs at a variety of campuses. You don't need to know which of these scenarios are right for you—yet. In chapter 4, you'll use a search strategy to discover what types of research are done on your campus and some options for exploring off-campus research adventures. But regardless of where your research experience takes place—at your home institution, off campus at a zoo or a hospital, or at another academic institution that becomes your home away from home, the best place is the one with a program that becomes a meaningful use of your time.

WET LAB AND DRY LAB RESEARCH

Some researchers use the terms *wet lab* and *dry lab* to refer to a type of research that is conducted, or to label a physical space where work is done. The distinction between wet and dry lab work is generally applicable to the life science disciplines. **A *wet laboratory*, or *wet lab*, is a space equipped with liquid chemicals and reagents.** The research conducted in a wet lab includes manipulation of samples through a variety of techniques that are done at a research bench with equipment such as a centrifuge, microscope, or other apparatus. **A *dry lab*, by contrast, is one where the work done might be primarily data analysis, computer modeling, bioinformatics, robotics, or other scientific pursuits that are typically done at a desk or table and don't involve liquids.** Many research groups conduct both wet and dry lab research. For example, scientists in a group that studies protein structure would prepare protein crystals in a wet lab, whereas their dry lab research might include analyzing the associated crystallographic data.

In our lab, we do both wet and dry lab research, but it so happens that they are done in the same room, although we've designated areas for each type of work. Our wet lab research is always done on the appropriate benches, and the dry lab research is done in a designated computer area. In our lab lingo, we don't distinguish between wet and dry work—we call it all research. But other research groups might do otherwise.

In this book, we use the word *lab* as to describe a group of people who conduct dry lab research, wet lab research, fieldwork (including work done at a field site or station or elsewhere), clinical research, or a combination of any of these. Depending on the context, we also use the term *lab* to refer to a physical space where research is conducted. Many scientists interchange the term *lab* to either refer to a group of people or a physical space, so you'll probably encounter this often as an undergrad researcher.

FIELD RESEARCH

Depending on your major, you might immediately think of fieldwork as interacting with or observing individual people or communities or observing animals and their actions in the wild. In any of those cases, you'd be correct. But *field research*, also referred to as *fieldwork*, encompasses an even greater and diverse range of methods and disciplines in the social, environmental, ecological, biological, geological, and other sciences. Sometimes, fieldwork opportunities are completed on or near a campus, and others are conducted at national or international sites. Field research might be seasonally dependent or independent.

Because fieldwork is not done in a laboratory environment, it might seem odd at first that we've included such work in the catchall *lab* for this book. But even when the data for a study are collected outside a laboratory, at some point they are brought out of a field site for analysis or discussion, or, in the case of actual samples collected during fieldwork, to categorize and properly store or to process in a wet or dry lab environment. These examples are only a subset of the vast array of fieldwork research experiences.

CLINICAL RESEARCH

Clinical research might take place in a lab, hospital, or dental clinic, or outside a medical establishment. It might involve direct interactions with patients or processing tissues collected from people, conducting surveys on a variety of topics, or other pursuits. **Perhaps the best-known type of clinical research is clinical trials**, which might test a new treatment for a disease directly on people to determine whether a method (or drug, or medical device, or other therapy) is overall both effective and safe. A clinical research study might include both wet lab and dry lab phases.

Sometimes, premedical students are given the impression (or told outright) that the only research experiences that will help them get into med school are clinical ones. This is unfortunate and untrue in our experience because most undergrad researchers who have done research in our lab have gone on to attend med school—an experience echoed by many of our colleagues who also don't conduct clinical research.

BASIC AND APPLIED RESEARCH

Scientific research can be categorized into two broad types—basic and applied. ***Basic research* focuses on understanding fundamental processes, whereas *applied research* is directed at solving a particular problem. In practice, there can be overlap between basic and applied research.** For example, someone conducting basic research to understand the regulation of protein synthesis in bacteria may discover a new target for antibiotics. Therefore, the knowledge gained from basic research may have obvious and immediate applications. Similarly, someone carrying out applied research aimed at curing a specific disease might uncover a new signaling pathway, thus adding new knowledge about cell signaling. (You may be aware of research that is called *translational* or *clinical*, both of which specifically address human health and are often considered to be applied research.)

MODEL ORGANISMS AND WHY THEY ARE SOMETIMES USED

Although the biological world is composed of countless organisms, researchers conducting basic research typically work with only a few. These organisms are known as *reference* or *model organisms*. Not all research groups use model organisms in their research, but some readers will consider research experiences involving them, so we're providing an overview here.

Model organisms are chosen for study based on life history traits that make them amenable for lab research. For example, a geneticist might study fruit flies (*Drosophila melanogaster*) not because they are particularly fond of fruit or flies but because fruit flies are easy to culture in the lab, have a short generation time, are easy to cross, and have a small number of chromosomes. However, if a scientist wants to study photosynthesis to understand the movement of electrons to create more efficient solar collectors, then fruit flies wouldn't work because they aren't photosynthetic organisms. A more appropriate model organism in this case would be the algae *Chlamydomonas reinhardtii*.

Research on model organisms is about understanding the unknown and the discovery of something new. Although this is not a complete list, other model systems scientists use in the lab include bacteria (*Escherichia* *coli*), yeast (*Saccharomyces cerevisiae*), plants (*Arabidopsis thaliana*), fish (*Danio rerio*), mice (*Mus musculus*), and worms (*Caenorhabditis elegans*).

DON'T JUDGE A LAB BY ITS MODEL ORGANISM

Over the years, we've connected with undergrads who were hesitant to work in a "mouse lab" because they didn't want to handle animals, and students who wanted to avoid a plant lab because they didn't want a botany project. Although these perspectives are understandable, they don't necessarily reflect the reality of the work done by a research group. For example, **you could conduct research for four years and never handle or observe the organism the lab uses.** This is because much research is done on pieces and parts such as tissue samples, proteins, or purified DNA. **In practice, a project could involve cell culture, protein expression, cloning, assays, or microscopy, and you'd never know, based on the techniques, if the model organism was mouse, fly, fish, plant, worm, or algae.** In addition, you'll miss out on incredible opportunities if you make an incorrect assumption about what kind of research is done on a model organism. For example, a research project that involves protein folding could lead to downstream applications that are important in human diseases such as frontal temporal dementia or cystic fibrosis. A scientist could study these in a lab that works with mice, human cell culture, worms, flies, yeast, or plants. Yes, that's right, plants.

Therefore, you shouldn't choose or rule out a research experience based on the lab's model organism, but instead base your search on whether the science inspires you. **Don't close yourself off to opportunities without knowing what they offer.** In most cases it's the science that you'll connect with—not the organism.

HOW PROFESSORS CLASSIFY THEIR RESEARCH PROGRAMS

Each professor uses a set of keywords and categories to describe their research program. Sometimes the categories or disciplines overlap with multiple fields (this is called *multidisciplinary*). For instance, a professor might use cell biology, genetics, microbiology, and bioinformatics to describe their research program or might simply use the term *cellular* or *molecular biology* to encompass all the above. This is important to remember when you're considering which research groups to apply to. You won't want to immediately discount a research opportunity because it doesn't list a specific topic or discipline you're interested in (because the professor doesn't list it as a specialty). Instead, you'll want to use their description to determine whether the research program is interesting or inspiring to you.

RESEARCH IS MORE THAN WET OR DRY BENCHWORK OR FIELDWORK

When you think about research, what comes to mind? Scientists in white coats wearing goggles or gloves? People staring through a microscope or at a computer screen? Perhaps a team of researchers excavating fossils? For many people whose only exposure to research is through movies or television, the initial response to "What is research?" is a variation on those themes. But even if a research project involves these, ultimately the research process is more complicated.

Although it varies by discipline and specific project, there are several broad, overlapping parts to managing a research project. Some parts, such as choosing how data are disseminated, are typically completed by the professor in whose lab the research is conducted. Other parts, such as updating lab notebooks or keeping a field journal, are basic expectations of all lab members. As an undergrad, if your project involves field research, you might help set up or test cameras or do other technical work starting on the first day. But if you're working in some disciplines on a project in a wet lab, you might not plan an experiment until you've learned background information and gained technical skills, even though you'll start taking notes on or soon after your first day.

NINE PARTS TO A RESEARCH PROJECT, SIMPLIFIED

No matter what research experience you join, it's important to understand that even a so-called basic research project is more complicated than you might initially be aware. Try to become involved in as many areas of the project management described next as possible during your research experience. For some parts, your participation won't be as in-depth as it would be for a grad student or postdoc, but as an undergrad, understanding what happens behind the scenes is still beneficial. **The parts of a research project described next presume that an undergrad researcher did not write the original research proposal and is instead working on a line of research that was funded before joining the group.** The sequence of the parts to a research project, and the responsibilities for someone who designs a study and secures funding for it, are slightly different.

PART 1: SECURING FUNDING FOR A SPECIFIC LINE OF RESEARCH

From supplies to salaries, research is expensive. In a research group, securing funding to support a project might be solely the professor's responsibility, or some professional researchers or grad students might partially or fully fund their position through grants or fellowships. At some institutions, undergrads can apply for funding to purchase lab or field supplies for an independent research project, but most professors don't require it. Many (if not most) undergrad projects are subprojects that are funded by a grant agency or other source secured by the student's principal investigator (PI) or in-lab mentor.

PART 2: LEARNING BACKGROUND RESEARCH

It's a given that whoever writes a research proposal is well versed in the background of the scientific field. But if you join project after a proposal is funded, you'll still need to learn the associated background information during your research experience. In essence, there are two categories of background research: project background and technical background. *Project background* includes the relevant information about the field (or subfield), basic information about a project, why it's important, and how it supports the overall research goals of the research group, when applicable. *Technical background* includes the purpose of the methods used in a project, knowledge of what happens during each step of a protocol or procedure, and how using a specific procedure helps achieve project objectives. The more background you understand, the more you'll connect with your research and be able to contribute to all parts of your research project.

PART 3: PREPARING TO DO THE WORK

After putting together a research study and securing funding for it as described in part 1, researchers break each large project goal into manageable tasks. This step entails gathering the necessary supplies and reagents, preparing the equipment, or traveling to the field site to conduct the research or retrieve samples or data to bring back to a lab. Depending on the complexity of the experiment and the materials available, this can take a few minutes, a few hours, or a few months.

PART 4: DOING THE WORK

These are the tasks that novice researchers imagine when they think about research—wearing gloves and goggles, possibly staring through a microscope or at a computer screen, and often working in a wet lab. But as we've covered previously, the type of work varies greatly depending on where it's carried out (a wet or dry lab, clinic, or field), academic discipline, and the procedures used. Field research, for example, might involve setting up traps to collect insects or setting up a portable flume system to measure water velocity in a sea grass community, whereas wet lab benchwork might include carrying out an experiment and redoing it if a procedure fails or gives inconclusive results.

PART 5: RECORDING NOTES IN A NOTEBOOK OR FIELD JOURNAL

If it wasn't written down, it didn't happen. Researchers spend a significant amount of time writing down what they plan to do, what they did, the mistakes they made in the process, and what they learned. In field research, someone might record notes to expand on and fill in greater detail later, methodical notes detailing a procedure, or personal reflections that can be used later to help prevent bias. In a wet lab, faithful recording of procedures, observations, results, and data is paramount because the results of successful experiment are unusable in grants and scientific publications if the details aren't correctly recorded in a notebook. Although it varies among research labs, some groups use sewn notebooks (similar to what you may have used in an instructional lab course), or an electronic lab notebook (whereby researchers upload their notes to a cloud-based repository), or a system of three-ring binders and computer paper, or college-ruled composition notebooks. After you join a research group, your mentor will let you know what system is used in that lab.

PART 6: ANALYZING RESULTS AND DATA

For many scientists, this is the most rewarding part of research. This is when discoveries are made, hypotheses are supported or nullified, and new models are derived. Analysis of new results and data lead to exciting new questions to ask, directions to take, and experiments to plan or lines of research inquires to pursue.

PART 7: PARTICIPATING IN SCIENCE COMMUNICATION

Sometimes undergrad researchers believe that they don't know enough about science to participate in science communication. But this is a myth. At its essence, science communication makes information about research accessible to nonexperts. Topics might include why your project is important; what tasks you do in the clinic, field, or lab; or your personal experiences about what it's like to conduct research. Whether these conversations take place online through social media or in person with friends, classmates, or family members, these interactions are science communication. In addition, science communication is sharing the results and other details of your research project in a thesis or a coauthored paper in a scientific journal, a peer-reviewed undergrad publication, or a poster presentation or talk at a conference.

PART 8: DELIVERING THE RESULTS

A scientist might produce data, results, or a product during their project, but that doesn't mean they own any part of it. Generally, for a federally funded grant to an academic institution in the United States, the results or products don't belong to an individual researcher or lab group but to the institution that administers the grant. Whether the details on how a project is completed and its results are shared in a scientific paper or thesis, or new a product is created and sent to an archive so other scientists can access them, researchers have an obligation, often a legal one, to share.

PART 9: PURSUING PRACTICAL APPLICATIONS

Sometimes the products created during a research project have practical applications in an agricultural, medical, or consumer marketplace. As an undergrad researcher, you won't be responsible for determining whether your research results are suitable for any of these—the professor whose lab you do research in will take the lead. However, it's entirely appropriate to ask if the professor is aware of any practical applications that might be associated with your research or suggest ones that you think are possible.

THE RESEARCH TEAM AND WORK ENVIRONMENT

A research environment can be an intimidating environment for new members—even when everyone is inviting. In this section, we introduce the roles of some researchers who might be part of the team you join and provide an overview of some aspects of lab culture that you might experience. However, it's best to consider this section as a guideline because research group members and cultures aren't identical across the board.

RESEARCH GROUP MEMBERS AND THEIR ROLES

The positions described in this section are generalizations of the roles and responsibilities held by some lab members in some research groups. Some grad students, for example, teach lecture or lab courses for undergrads throughout their degree program whereas others might only teach for one year or not at all. And some responsibilities, such as ordering supplies, might be an expectation for all group members, or the task might be assigned to a single person. Details on how the makeup of the research group might or might not influence your research experience are covered in chapter 6.

The Principal Investigator (PI)

Note: The abbreviation *PI* will be used extensively throughout this book.

The PI is the overall lab head. The title you're likely most familiar with is professor. The PI is an expert in their field of research, has earned a PhD (or other advanced degree such as an MD), and typically had several years of experience as a professional researcher before starting a faculty position. Sometimes PIs don't teach classes during the summer and instead focus on their research program. A PI's research responsibilities include determining the research priorities and overall goals of the research team, obtaining funding, publishing results, and making personnel decisions. A PI often teaches courses, writes recommendation letters for research group members (and students from classes), and participates in various committees in their home department and at the college or university levels. Additional responsibilities may exist depending on the academic discipline and the college or university to which the PI belongs.

Professional Research Staff (PRS)

A member of the PRS could hold any number of positions or roles within a research group and might work full-time or part-time—it varies depending on the researcher's appointment and the person's career goals. Individuals in this category might include technicians, research assistants, research scientists, and others. Degree type also varies. A PRS member might have earned a bachelor's degree, or a bachelor's degree and a master's degree, or a bachelor's degree, then a master's degree, and then a PhD degree. A member may stay in the same research group for only a few months or might be employed in the PI's lab for several years. If there is a long-established PRS person in the research group, they are probably the official lab manager.

POSTDOC RESEARCHER

Postdocs hold a PhD degree, thus the abbreviation *postdoc* for postdoctoral. A postdoc's responsibilities can vary significantly among labs, or even within the same lab. If no one officially has been appointed to the position of lab manager, a postdoc might adopt the role. A postdoc's appointment might be temporary, with goals to publish papers, gain research experience, and develop a research program of their own prior to applying for faculty positions, or it may be longer-term as described for PRS. Depending on their official responsibilities, outside their research obligations, a postdoc might teach lecture classes, write grant proposals and manuscripts, create posters, present research at scientific meetings, and serve on grad committees.

OTHER PAID RESEARCHERS

A research group might include several employees who contribute to the success of the research program in a variety of positions and roles. **Undergrad students with paid research positions are often included in this group**, as are employees who have some research experience but haven't been in the workforce long and scholars who are visiting from other institutions. Some researchers who aren't hired on a salary contract might also fit in this category. There are other possibilities as well.

Student Researchers

Students at various levels of their education may participate in research, typically for academic credit or experience rather than for pay.

UNDERGRAD

Essentially, **there is no one-size-fits-all undergrad research experience.** Even students who work on similar projects with the same mentor will have unique research experiences because many aspects become customized for an individual based on a variety of influences. Some of these influences might be a student's immediate and longer-term academic goals, their dedication to a project, time available to conduct research, and determination or resilience when navigating boring or frustrating research moments. Within this framework, how a student values their experience (for example, whether they consider their time spent in the lab as a checkmark for their resume or a valuable learning opportunity) and the actions they take to demonstrate that belief will influence their daily responsibilities and the development of mentoring relationships with labmates.

There are also numerous potential administrative-like variables that might have an impact on an undergrad's research experience. For instance, in some research groups, an undergrad will define their own research question, whereas in others their mentor assigns independent projects to student researchers. In other research groups, an undergrad will primarily assist with another labmate's project. Furthermore, some undergrads participate in research part-time during the regular semester, pay tuition, and receive academic credit for their efforts. Other undergrads participate in paid research experiences exclusively during the summer months. (These distinctions are covered in greater detail in the "Lab Culture" section later in this chapter and in the chapter 3 section "Understanding and Managing Your Expectations.")

GRAD

A grad student's research responsibilities include substantial benchwork or fieldwork with the goal of completing a thesis or dissertation (unless pursuing a nonthesis degree), learning research techniques and procedures, and preparing as much as possible for their next career phase. A grad student might be pursuing either a master's or doctoral degree. A grad student might mentor undergrad students or have responsibilities that include aspects of lab management such as ordering supplies, making stocks, and helping to maintain equipment; some perform both functions. Outside the lab, a grad student takes classes at the beginning of their grad career and might serve as a teaching assistant throughout it. They are also typically responsible for writing manuscripts on their research and presenting their work at scientific conferences.

HIGH SCHOOL

Some programs match high school students in college and university labs to participate in a research experience. Typically, these programs are done over the summer months and are an opportunity for PIs to give back to the community, as well as provide an inspiring research experience for the student. In some programs, the students are paid a stipend through a fellowship, and in others the student pays to participate. (PIs typically don't financially benefit from hosting high school students even if the student pays to participate in their research program, but there are exceptions.) In other scenarios, a hybrid high school and university program allows students be part of a professional research program during the school year.

BEHIND THE SCENES: WHAT THE NONUNDERGRAD RESEARCH GROUP MEMBERS DO ALL DAY

A research lab is a professional work environment. As an undergrad, you're a part of it, but you aren't subject to the same stressors, pressure, responsibilities, or commitments that affect the other members of the group. For example: Weekends? After 7:00 p.m.? Official holidays? You might use these as opportunities to get caught up on sleep, take a road trip, go home for a few days, finish a class assignment, or work extra hours at a job. For a professional researcher or grad student, these might be opportunities to get more research done without the responsibilities of classes, supervisory duties, or campus parking issues. In an academic setting, ideally your labmates will arrange their workdays to coexist with their personal life and professional development goals, which might mean, in part, working outside of a traditional nine-to-five workday or at times when you wouldn't be around.

Although it varies by person and discipline, some tasks that can be found on the agenda for a professional researcher or grad student might be the following:

- Research. Research. Research. This might include wet, dry, field, clinical, or a combination of these. It's likely that your labmates will conduct research or do research-related tasks for more hours each day than you will do per week. That's okay—you're in a different part of your career than they are, and you have different levels of responsibilities! (An exception might be if you're participating in a full-time summer research experience; then you might match hours per week with some of your labmates for that period.)
- Designing, implementing, and performing data analysis of experiments or research studies
- Determining when to abandon an experiment or redesign it
- Planning and assisting on undergrad (or grad student) projects
- Preparing stocks and reagents—the ones on the benches, in the cabinets, in the freezers, and in the refrigerators
- Reporting their results and data to the PI or giving general progress updates or plans for new directions
- Writing journal articles, grant proposals, fellowship applications, peer-reviews of other scientists' papers, and possibly a thesis or dissertation
- Preparing to present at a scientific conference. Traveling to said conference. Stressing about the work to be done because they are at a scientific conference instead of doing research.
- Reading peer-reviewed journal articles
- Troubleshooting undergrad experiments, their own experiments, other lab members' experiments, and the experiments of researchers from labs down the hall or across the world
- Making coffee. Drinking coffee. Thinking about making coffee. And drinking coffee. (Or tea, or whichever caffeinated beverage they prefer if they consume caffeine.)
- Working on various aspects of professional development outside the lab
- Participating in or starting outreach programs or science communication on social media
- Trying methods to incorporate self-care and wellness into their daily routine and striving to achieve a life and research balance that works for them

YOUR MENTORS

Although it's impossible to characterize every mentor's role for every undergrad researcher, a mentor will offer instruction, training, and encouragement, although the levels vary significantly depending on the individuals involved. A mentor might also provide suggestions about opportunities that will enhance professional development and will care about an undergrad's ultimate goal after graduation. A mentor might suggest ways to use a college experience to accomplish personal goals and encourage each student to make a full investment in their college experience. In some instances, a mentor might connect an undergrad to campus resources to help them manage a health-related matter or personal issue.

Depending on the research group you join, you might form connections with several mentors, advisors, or role models during your research experience. For example, you might have a single person who works closely with you on your project (an in-lab mentor), another person who provides mentorship in various issues but doesn't direct your research (a labmate or two), and the PI who mentors you in several matters. However, an individual won't necessarily fit into such distinct categories. A PI, for example, might both work with you directly on your research project and mentor you in professional development or other categories.

The positive impact mentorship can make on an undergrad researcher is immense—often the full effect isn't realized until a few years after graduation. This is especially true for those who start a research experience early and have the benefit of a mentoring relationship throughout their undergrad career.

In this book, we use the term *in-lab mentor* to refer to the person who directs the majority of your training and research project. Your in-lab mentor might be the person who conducts your interview and invites you to join the research group, be instructed by the PI to work with you, or be someone who volunteers to work with you after you've been in the lab for a while. This person might be a grad student, a postdoc, another member of the PRS, another undergrad, or the PI. Who your in-lab mentor will be is influenced, in part, by the type of academic institution you attend and the training culture of the research group you join.

In labs where an in-lab mentor isn't assigned at the start of a research experience, which happens sometimes in wet or dry labs but is less likely in field research, a student might learn how to follow a procedure from one person, learn to make a solution from another, and learn to keep a proper notebook from yet another. This can also be the case for students who design their own research project. When an in-lab mentor is determined before an interview is scheduled, a new student typically has a single, go-to person starting from their first day. However, we hope that during your undergrad experience you'll develop professional mentoring relationships with multiple group members even though a single person will likely be responsible for directing your project.

Most likely, on paper, the PI will be considered your official mentor, but they might not be your in-lab mentor. This can be confusing for new undergrad researchers—especially if they interviewed with the PI. In addition, students enrolled in degree programs at teaching-focused institutions but are participating in a research experience at a research-intensive institution might be surprised if the PI doesn't work closely with them during each session (or at all). In every case, it's important to understand that the contact hours a PI has with research group members are related to the lab culture and the PI's management or mentoring style.

The type of interactions a PI has with each group member depends on a variety of factors but is strongly influenced by each researcher's training needs, the makeup of the lab team, and the PI's mentoring and training strategies. **In some labs, the PI meets with undergrad researchers each lab session** (or several times a week) and might teach protocols or techniques, or they might conduct fieldwork together. Through these interactions, a mentoring partnership generally is created between themself and each student.

In other research groups, a PI might not directly work with undergrad researchers (or any of their research team) in the lab or the field. In these cases, a PI most likely instructs other group members to train and mentor the undergrads. Even in all-undergrad labs, a PI might instruct the more experienced students to train new members in the core research techniques or how to use equipment. In some labs the undergrad researchers aren't paired with an in-lab mentor at all but are expected to use self-directed learning for much of their research experience.

On the surface, none of these approaches are better nor worse than the others, although generally we don't believe that an undergrad research experience is as beneficial for students who do most of their own training. But overall, what matters most is what is best for you: if you receive the assistance you need to be successful with your project; if you're working in a safe, respectful, and inclusive environment; and if the opportunities for professional and interpersonal development and building your mentoring network are accessible through your research experience.

Even if you don't work directly with your PI on a research project, most likely, as you demonstrate reliability and meet with your PI to discuss your research and career goals, you'll build a mentoring relationship. However, until then, know that the PI makes an immense impact on your research experience and training simply because you're a member of the research team.

LAB CULTURE

A lab's *culture* is simply this: a minisociety consisting of a group of researchers, a somewhat flexible system of rules and etiquette, and a basic understanding (although not necessarily complete agreement) among members on how the lab should function. **Embedded in a lab's culture is everything and everyone connected to how the lab operates and how its members interact with one another.** This is why a research culture is a complicated entity. Think of it as a living, evolving organism. In some research groups, the culture is so fluid that the addition or loss of a single member can cause a substantial change, which can result in a positive or negative shift.

Specifically, some aspects of lab culture include the PI's management style, the system for training personnel (including undergrads), how carefully safety regulations are followed, how equitably lab chores are distributed among group members, whether each researcher has a personal research space, how supplies and reagents are ordered and shared, whether labmates freely share their wisdom and expertise to help one another succeed, whether group members are supportive of one another in times of failure, and how much undergrad researchers are appreciated and valued. Within these categories are seemingly infinite variables, which is why no one-size-fits-all lab culture exists. Therefore, what is true in one research group isn't necessarily true in another.

HOW LAB CULTURE IS ESTABLISHED

Essentially, there are three significant influences on lab culture: (1) the PI's management style, (2) the personalities of the individuals in the research group, and (3) the work habits of each person. These influences are intertwined, but when a PI insists on a healthy and respectful learning environment, then typically members of the research team create one.

WHAT YOU NEED TO KNOW ABOUT THE PI'S MANAGEMENT STYLE BEFORE CHOOSING A RESEARCH EXPERIENCE

Sometimes an undergrad is advised to ask a PI to explain their management style before accepting a research position. However, this opportunity won't be readily available if you're not interviewing with the PI but instead with a grad student or member of the PRS. And if you're not going to work directly with the PI, but instead with another member of the research group, the PI's management style might not have a strong impact on your overall research experience. But in a short interview, even when discussing it directly with the PI, it can be difficult to fully understand their management style and what impact it might have on your research experience as an undergrad.[1] (For a grad student or member of the PRS, the PI's management style is absolutely critical.) **When asked, most PIs will gladly explain a combination of their management style and mentoring philosophy with some aspects of the lab culture thrown in.** But because you'll interpret that explanation through your own filter and possibly without the knowledge of what it's like to work in a professional lab environment, you won't have the tools to completely evaluate their approaches. For example, if they state that they are *always* available when students have questions, does that mean you can call, text, or email at any time and expect an immediate response? (Maybe, but probably not.) Or does it mean that they will be available but not necessarily right away, and sometimes it might take a few hours or a couple of days before you receive a response? (This is more likely.)

However, if they state that instead of solving research problems for students, they expect undergrads to find the answers they need, does this mean that they will coach you during conversations until you arrive at the correct answer or that you'll be on your own to find answers from the internet and labmates? (There are PIs who are fit squarely into one of those categories and PIs who move between the two depending on the moment and what they assess would be most beneficial for the student or what their immediate schedule allows for.) We're not suggesting that you shouldn't ask a PI about their management style, just know that their answer might or might not be enlightening.

Therefore, although we won't discourage you from doing it, inquiring

1. One exception to this to might be if a current or former member of the research group warns you that the PI (or the potential in-lab mentor you interviewed with) is inappropriate in their behavior toward the members of their research team. Another exception would be if you're in an interview and are made to feel unintelligent, fearful, or uncomfortable. There is a difference between feeling somewhat doubtful or self-conscious at an interview (which is common) and feeling intimated or unwelcome by the interviewer (which can be a sign of an unhealthy workplace). Another "red flag" would be if the interviewer is openly hostile to who you are—even if they try to pretend something they say is a joke. An interviewer who expresses racism, sexism, ableism, transphobia, or other discriminatory perspectives will likely continue to do so in the future. If you experience this sort of behavior by the interviewer, it's likely that the research group and mentor are ones you should avoid.

about a PI's management style at an interview probably won't be as informative as asking questions about the types of techniques you'll learn or the specific responsibilities or opportunities you'll receive as a group member. Questions to ask in an interview with a PI or another potential in-lab mentor are covered in chapter 6. For most undergrads, the PI's management style won't be an issue. But if it does turn out to be problematic, you can take your transferable skills to a new research experience.

LAB CULTURE AND YOUR UNDERGRAD RESEARCH EXPERIENCE

Even though lab culture is connected to everyone and everything in a research experience, some areas will have a more substantial impact on your undergrad research experience than others. For example, the application procedure won't matter much, but the commitment level will. And the technical training opportunities available will matter significantly *if* they are connected to achieving your goals. And whether you're a member of an underrepresented group in STEMM, you'll absolutely benefit from joining a culture where inclusivity, diversity, equitability, and accessibility for researchers with disabilities are genuinely valued by all team members. Next, we've outlined some aspects of lab culture that affect undergrads and how might they differ among research groups and/or mentors.

SEVENTEEN WAYS CULTURES FOR UNDERGRADS MAY DIFFER AMONG RESEARCH GROUPS

1. **Experience requirement.** What counts as experience and how much, if any, is required before joining a research group varies. Not all mentors require experience, and some prefer undergrads who don't have any. We cover this topic in more detail in chapter 3, in the section "The Experience Paradox."
2. **Application procedure.** There isn't a standardized procedure to apply for a research position. Each potential mentor determines the selection criteria, application process, and interview procedure that works for them. Some use a quick selection process, and others follow a substantial vetting strategy that includes volunteering as a member of the research group before being offered a research spot. If you're applying to several full-time summer research programs, the applications among institutions will be similar but not identical. We discuss strategies for applying to undergrad research positions in depth in chapter 5.
3. **Time commitment and schedule.** To join a specific research project, most mentors require a minimum time commitment in hours per week

and semesters. How flexible that commitment can be might or might not be negotiable. This topic is discussed in more detail in chapter 3 in the section "The Time Commitment."

4. **Types of positions.** How much time an undergrad spends observing what others do; performing benchwork, clinical work, or fieldwork; or completing research-related tasks such as washing glassware or reading journal articles differs vastly among research groups. For some undergrads, an entire experience might only include observing labmates or doing labkeeping chores and not involve any actual research. For others, the core of their experience will be an independently conducted project starting on the first day. A longer explanation of these possibilities is presented in chapter 3, in the section "Wet Lab Positions for Undergrads."
5. **In-person or virtual projects.** In the spring of 2020, when the COVID-19 pandemic became a force that pushed colleges and universities to take their classes online, research experiences for undergrads were also affected. Over the next few months, some undergrad research programs were outright canceled, but others were moved into the virtual realm. Consequently, a wider acceptance of distance-learning research opportunities for undergrads grew. Now, in some research groups, an entire project might be conducted virtually, but in others a project still must be completed in a lab or clinic or at a research site. However, some mentors now offer projects that have both virtual and in-person components.
6. **Determination to foster an inclusive, diverse, equitable, and accessible learning environment.** Many research groups are supportive and welcoming to students from underrepresented and underserved backgrounds in STEMM. In these research groups, the PI and other members also strive to identify their own implicit biases that could, among other issues, negatively impact the personal or professional development of their coworkers, including the students on the team. However, unfortunately, there are individuals who don't value diversity, inclusion, equity, and accessibility for students with disabilities and may participate in hurtful and inexcusable behavior through racist, sexist, ableist, anti-LGBTQIA+, or other discriminatory actions toward their labmates. Although we hope that all members of the research group you join respect everyone else and behave accordingly, we don't have a surefire strategy to avoid those who don't unless someone demonstrates inappropriate or egregious behavior during an interview.
7. **In-lab mentor's training style.** Each in-lab mentor determines the training style that works best for them. Sometimes that training style matches an undergrad's expectations, and sometimes it doesn't. Some mentors meet daily or weekly with an undergrad researcher, and others are only available a few times during the semester. Some mentors are available

after 5:00 p.m. or on weekends and will answer emails, texts, or phone calls quickly, and others basically disappear into the ether after the interview. Some mentors will guide a student through a modeling program step-by-step, and others will indicate the pH meter and instruct the student to figure it out. Even in the same research group, a mentor might distribute their time disproportionately among all the undergrad members.

8. **Technical training opportunities.** The opportunities a research experience provides in this category are extremely important to ask about in an interview and to keep in mind when deciding whether to accept a specific research position. Certain projects require learning a set of standard techniques before conducting an experiment or collecting data. Other projects require becoming an expert in a single technique and then using it to do discovery-based research. Some projects will be a combination of wet and dry benchwork or fieldwork, and others will fit squarely into one of those categories. A project might include techniques from a single discipline or encompass a broad range of cross-disciplines. These are only a few possibilities.
9. **Project contribution.** Some PIs or in-lab mentors expect undergrads to participate fully in a project from start to finish, including the initial design strategy, the actual research, data analysis, figuring out the next research question to ask, and writing up the results in some form. Other PIs direct undergrads to assist experienced researchers on subprojects or work on different aspects of several projects. There are additional scenarios not covered here.
10. **Requirements for understanding the project background and applying classroom knowledge to research.** How much will be expected in this area will depend on how important it is to the PI or in-lab mentor. Some PIs believe that the type and amount of knowledge demonstrated by an undergrad is directly linked to how motivated and creative a student is, and these PIs mention so in recommendation letters. Other PIs and mentors believe that it's less important for an undergrad to take a holistic approach to their research experience as long as a solid understanding of the study and the overall research goals of the research program are demonstrated. Some PIs and in-lab mentors don't have strong expectations either way provided the research gets done.
11. **Peer-training opportunities.** Some PIs believe that it's important for undergrads to help train their peers in basic techniques or procedures and ensure that these leadership opportunities are available whenever possible. Other PIs prefer that all training is carried out by themselves, grad students, or a member of the PRS.
12. **Responsibilities outside the lab.** In some research experiences, all work is completed during regularly scheduled research hours. In others, work

outside those hours can be substantial. For example, your mentor (or department) may require writing a research proposal, end-of-semester report, or short paper; constructing a poster; attending a symposium; reading several papers written by the research group; or attending group meetings. Scheduling time to complete these tasks during your research hours may not be possible in your research experience. Some undergrads are also required to enroll in a seminar or workshop-based class before participating in research or after they begin working with a research group. Typically, these are one-credit classes that meet once a week. (You may be reading this book because it's a required text for one of those classes.)

13. **Course credit.** Whether a student is able, allowed, or required to register for undergrad research varies but is mostly tied to the PI's preference and the policies of the department where the research is conducted. Some PIs require registering for credit during the first semester, and other PIs only allow it after a student has demonstrated genuine interest in the research program and acquired a certain skill set. In some instances, even if the PI doesn't have a preference, students conducting research are required to register for course credit by their home department or if they wish to receive honors credit for their efforts. A longer discussion on course credit for participating in undergrad research is presented in chapter 3, in the section "Deciding Whether to Register for Course Credit for Research."
14. **Salary.** Some PIs don't pay undergrad researchers to work on a research project, and others pay undergrads to work on a specific project or as a general lab assistant. If an undergrad is part of a summer research program, a stipend and room-and-board allowance are often awarded, but some programs require the student to pay to participate. In chapter 3, details on securing a paid summer research experience are presented in the section "Summer Undergrad Research Programs (SURPs)," and possibilities for students conducting research during the semester or quarter system are covered in the section "Wet Lab Positions for Undergrads."
15. **Undergrads as coauthors on publications.** In many research groups, the PI determines authorships for publications, although sometimes a grad student or member of the PRS is charged with the responsibility. When they have earned the recognition, most PIs include undergrads on peer-reviewed publications. The same goes for posters or abstracts that aren't presented or written by the undergrad but where the student contributed to the research presented. (For a variety of reasons, not all research results and data end up in a publication or presentation.) Some mentors also encourage students who wish to publish their research to write a senior thesis or a paper for an undergrad research journal.
16. **Conference attendance.** The emphasis placed on attending conferences (such as an on-campus symposium or scientific meeting elsewhere) is

directly related to how much a mentor believes a particular student will benefit from the experience, the cost of attending, and if funds are available to send a student. Typically, a PI or an in-lab mentor (or both) will encourage an undergrad to attend meetings if they believe the student will benefit, contribute, and be a consummate professional at the event. Other PIs or in-lab mentors don't believe that large conferences at international meetings benefit most undergrad students and don't encourage attendance, although they don't technically discourage it, either. Finally, some PIs and in-lab mentors are directly involved in encouraging students to attend meetings and helping them prepare an agenda to get the most out of the event.

17. **Travel awards to conferences.** If an undergrad has enough results or data to present at a scientific conference and is likely to use the meeting to enhance their professional and personal development, most mentors will encourage the student to apply for travel awards either at the college level or from conference organizers. Some PIs will provide an additional travel stipend to help defray the cost of attending a conference.

3

Will I Like Conducting Research?

UNDERSTANDING AND MANAGING YOUR EXPECTATIONS—YOUR STRATEGY FOR HAPPINESS AND SUCCESS

Although there are no absolute guarantees that you'll like conducting research on a particular topic with a particular research group, **your enjoyment level will be influenced by what you want to gain from the experience, what you expect it to be like, and the research opportunity you ultimately join.** Several years ago, I (DGO) interviewed an undergrad who was unhappy with their research experience and in search of a new one. They were smart, hardworking, and genuinely interested in conducting research, yet they were unhappy with their research experience because they had "only learned some techniques and wasn't really doing any research."

But when they began to explain their project's responsibilities, I knew immediately that they were indeed conducting research. They were building the research tools needed to conduct an experiment and test a hypothesis. For them, a simple but unfortunate misunderstanding was the stumbling block to happiness: their expectations were very different from the reality of the research experience they chose. To have been happy in their first experience, they either needed a project that included hypothesis testing from the start of their experience, or they needed to understand how their daily activities were directly connected to the research program's big picture.

By the time they interviewed with me, the student was overwhelmed by frustration and was almost ready to give up research completely. This one example highlights why realistic expectations from the start help prevent disappointment later. Throughout this chapter, we present specific questions and answers to help you ultimately address this chapter's title question for yourself, equipped with a realistic set of expectations.

What influences my expectations and how do I know if I have any?

You have expectations for your research experience even if you're not consciously aware of what they are. Expectations include the goals you wish to achieve through a research experience and what you imagine it will be like to conduct research as a member of a research team. Your expectations are influenced by classmates' stories about their research experiences, professors' and teaching assistants' opinions, television shows and movies, and researchers and scientists you follow on social media. And, to some extent, your expectations are influenced by what you've read in this book so far.

Expectations can be so subtle that you're unaware of them, or so bold that it's nearly impossible to satisfy them. Either way, if you strive to understand what your expectations are, and if they are realistic, you'll be more likely to find a project that matches them. This approach will help you achieve your research and professional goals, which will make your experience a more meaningful and rewarding use of your time. **Ultimately, you'll be happier and more successful if your expectations of what research is closely match the experience you join.**

What are some examples of expectations that undergrads have?

Sometimes students, such as the one mentioned earlier, start a research experience with expectations that lead to substantial disappointment. Through our own mentoring experiences and discussions with researchers in several science fields, we know that sometimes a student's expectations for their research experience turn out to be vastly different from the reality of it. Some common expectations and the reality in this category follow.

Expectation: "I'm already an expert in the techniques from lab class, so research will be easy."

Reality: For most undergrads, research turns out to be more difficult and more complicated than anticipated—even after completing lab classes in the same discipline. It takes time, practice, personal resilience, the willingness to admit to mistakes, and the determination to use constructive criticism as a launchpad for future success to gain an arsenal of research skills. Even then, not every research technique will be easy to do because some are finicky, annoying, or boring. (We cover this situation in greater depth later in this chapter in the section "Using Lab Classes to Prepare for Undergrad Research.")

Expectation: "I'm going to do research with the same professor who taught my lab class, so I already know what to expect in their research lab."

Reality: A professor or a teaching assistant (TA) instructing in a lab course focuses on the associated responsibilities for the duration of the class. A professor who supervises a professional lab (or a TA working in a professional lab) has multiple, ongoing activities that require their attention during their workday. These differences in responsibilities necessitate that the two experiences won't be identical for you as a student. And it's possible that you won't work with your mentor during every research lab session, as discussed in chapter 2.

Expectation: "I only need research experience to get a recommendation letter, so the project doesn't matter."

Reality: The research experience you join does matter. To be happy and successful during yours, select a project that aligns with your interests or values, or you might soon find that trying to muster enough passion or enthusiasm for the project is more work than the research itself.

If you're only interested in listing "participated in research" on your resume or approaching the experience as a checkmark before asking for a recommendation letter, you'll set yourself up for disappointment. Being genuinely interested in the discipline or type of research, or techniques you'll learn, or why the study is important to conduct, makes it much easier to maintain the motivation to produce high-quality work, and it will be epically hard to do this for anything beyond a short time if you don't care about it. Plus, recommendation letters aren't part of a transaction for spending time conducting research, and some principal investigators (PIs) write letters only for undergrads who fully invest themselves in their research project. We cover more on this topic in chapter 4, in the section "Step 2: Identify Potential Meaningful Research Experiences."

Expectation: "I want to be passionate about *something*, so I picked research."

Reality: Passion can be ignited during a search, or it might be delayed until someone has acquired certain research skills to contribute to their project and enough knowledge to understand it. For some, it might not happen at all if the student and the project aren't a good match. That is why the search and interview processes are so important.

Expectation: "I have a friend who loves their research project, so I want to do it, too."

Reality: Each research experience is unique because each undergrad researcher and the professional relationship they develop with their in-lab mentor is unique. Even when two undergrads are members of the same research group and have similar projects, each student will have a customized experience based on their in-lab mentor, goals, and the personal investment they each make in their project. Having a friend who loves their research project can be inspirational and can guide you toward research, but your experience won't be an exact copy of theirs. It will be your own. Tips for asking classmates about their research experiences are presented in chapter 4.

Expectation: "Participating in research will make me feel important."

Reality: We hope so because the research you're doing is important. But it does take time to feel confident when working on a research project. At the start of a research experience, most undergrads feel awkward and experience more self-doubt than self-confidence. Undergrads who push through this phase are often rewarded with a transferable skill set and greater self-confidence, but until the transition is complete, it's common to feel more doubtful than important.

Expectation: "I'll do research for one semester so I can earn a publication to list on my med school applications."

Reality: Publications aren't guaranteed—no matter how hard you work. In many fields, even postdocs and grad students don't earn a publication in a scientific journal every semester. For some, it's even more likely to take three or four years to earn a first-author one. However, it's possible that you'll join the right project at the right time and produce publication-worthy data that result in a coauthorship in a single semester. Other possibilities might include publishing your work in an undergrad research journal or writing a senior thesis. If earning a publication is important to you, bring it up at the interview and make an appointment with your PI to discuss the possibility again after you join the research group. But definitely don't *count* on earning a publication in one semester.

Expectation: "I'll be cloning dogs and curing cancer."

Reality: Well, you might be, but probably not. The timeline between thrilling discoveries and exciting breakthroughs can be few and far between and in some disciplines; indeed, an individual scientist might only experience a few during their entire career. Research is mostly preparing to do a procedure (which is often a multistep process, perhaps by gathering samples or learning a skill set or reading journal articles) and then grinding through the actual process in a lab, a clinic, or at a field site.

Depending on the discipline, the next phase might entail cleaning up your mess and doing it all over again. And again. Plus, there are various research-related tasks that are part of a research experience, such as keeping a proper notebook, being a good labmate by doing some group chores, reading more journal articles, contacting potential collaborators to ask for assistance or access to their research samples, and presenting the results. For a refresher on the topic, revisit the section "Nine Parts to a Research Project, Simplified," in chapter 2.

Expectation: "I'll learn techniques during the first semester and design my own project and experiments in the second."

Reality: Some students design their project and the accompanying experiments or procedures. Sometimes they have a project in mind before joining a research group, but other times the PI requires a semester or more of working with an in-lab mentor before supporting this type of independent research.

However, in some disciplines or research groups, undergrads rarely—or never—design experiments or projects. In these cases, it might be because a project was already fully designed before they joined the lab (summer-only research projects in wet labs often work this way) or that there is virtually no chance that someone without several years of training and an extensive research skill set will be able to design an entire research project in a particular research group (this includes technical expertise and the ability to design effective controls).

But even in research groups where undergrads don't design a project, students contribute by completing experiments and procedures and making daily decisions on troubleshooting and workflow, interpreting data and results, and more. This system provides the benefits of working on an independent research project even if it was designed by the PI or an in-lab mentor. We will return to this topic later in this chapter.

Expectation: "I'm paying for course credit to do research, so it's reasonable to expect assistance when I want it."

Reality: Everybody waits in research. We might wait for our samples to incubate, spin, dry, or dissolve. We might wait to use equipment or for our mentor's assistance. Plus, all grad and professional researchers and PIs wait for papers and grants to be reviewed and then accepted or rejected, and for a collaborator or colleague to respond to an email.

Unless you have an immediate safety concern, be prepared to wait for a labmate's assistance now and again. It's reasonable to expect help in an appropriate time frame, but it's unrealistic to expect your labmates to immediately pause their research or immediately answer an email or text every time you require assistance.

Expectation: "I'll apply what I discover in research to my real-life problems and situations."

Reality: This is fair to expect if you join a research project that is applicable to your life outside the lab. But if you don't, then it's not a realistic expectation.

If you want to ensure that you're working on a project that explores real-life problems you connect with, use that goal to guide your search for research positions. Also note that it's acceptable to ask about how (or if) a project will help you fulfill this aspiration in an interview. We hope, though, that the interpersonal and professional development you'll gain from participating in undergrad research will add value to your life no matter the subject of your research project.

The take-home message on expectations is this: Having them for your research experience isn't a problem—in reality it's good. After all, it's those expectations that will guide you to a meaningful research experience. The key is to recognize what you believe conducting research will be like and reconcile it (as needed) with what the research experience you choose will *actually* be like. You'll do both during your search for a research position and by asking pointed questions at your research interviews. To help you prepare, this chapter covers common concerns and expectations students have before starting an undergrad research experience. **Essentially, this chapter includes the information that we wish we had known as undergrads before searching for our first research positions. And what we wish every undergrad was aware of before joining our research team.**

As your read the remainder of this chapter, keep in mind that a research experience, even one that is a meaningful use of your time, won't be without challenges. For example, you may feel awkward when first getting to know your labmates, frustrated when you're learning a technique, or bored with performing certain tasks related to your project. And no matter how interesting the topic, conducting research won't be fun or even personally rewarding every day. But overall, the experience you join should be worth the time you exchange elsewhere in your life to participate.

QUALIFICATIONS FOR PARTICIPATION IN UNDERGRAD RESEARCH

Sometimes students disqualify themselves from participating in undergrad research because they believe (or someone has told them) that their grade point average (GPA) isn't high enough, they aren't smart enough, they don't have enough prior research experience, or some other reason. But here's the truth: **You're in college, so you're qualified to do undergrad research.** You didn't trick an admissions committee to let you in, nor were you accepted to college because someone felt sorry for you. **You earned your place in college and you belong in a STEMM research experi-**

ence if that's what you want to do with your time. And you don't need to have a scientist in your family or be friends with one to be successful in undergrad research. We want to underscore this point for those readers who haven't received strong mentorship in this area and those from communities that are underrepresented in science, technology, engineering, math, and medicine.

For any student, you might not have the eligibility requirements (such as academic year) to be eligible for *certain* undergrad research positions, but you're qualified to conduct research and make a contribution to science in the process.

In the next section, we address the most common concerns on this topic we've encountered as mentors in interviews with undergrads, through social media, and in discussions with colleagues. **Our goal with this section is to prevent self-doubting students from preemptively disqualifying themselves from a research position before they've even started their search.** And if you're having doubts about something we didn't cover here, find us on Twitter @YouInTheLab or Instagram @UndergradInTheLab and send us a direct message detailing your concern. If we don't have the answer, we can send your inquiry anonymously out to the Hive to ask for suggestions.

Isn't being a genius (or at least brilliant) a requirement for doing undergrad research?

Can you work hard? Can you follow instructions? Will you be determined to learn from your labmates? When a research procedure fails (and that's *when*, not if), will you try to solve the problem rather than immediately giving up? If so, you've got this.

Most scientists aren't geniuses or even brilliant. What they are is interested in their research and resilient when managing disappointment and failure. They ask others for help when they need it, and when they make a mistake, they are open to learning how they can prevent a similar one in the future. If you share these characteristics, and you want to participate in undergrad research, then don't let anyone—including yourself—discourage you from pursuing a research experience because *you can do this.*

My friend was asked by a professor to join their research group, but no one asked me to become a research assistant. Does that make me less qualified for research?

Nope. Some professors (or grad teaching assistants) are always open to identifying inquisitive undergrads who might want to join their research team. Others don't think much about recruiting new students even though they are dedicated mentors. Your friend connected with a professor in the former category, which is great for them, but it bears no impact on your chances of landing a research position nor indicates that you're less qualified to conduct research. I (DGO) have recruited undergrads from lecture classes, and some grad TAs have done the same. But other undergrads have joined my research program after responding to an advertisement for a research position, asking about the possibility during office hours, or making an appointment after learning about my research program on their own.

I didn't grow up "speaking science." How will I be successful in a research position?

In interviews for undergrad research positions, questions about the difficulty of learning the new lab language routinely come up—often from first-generation college students. Some students are fearful that they will finally start a research experience and then be sidetracked by struggling with a language learning curve. I (PHG) remember a student who was concerned about understanding the research procedures because their parents did not use English at home and, as they said, "I didn't grow up with the lab terms, so I'm worried about learning them." I was glad that they mentioned this because it allowed me to immediately address why I knew it wouldn't be a barrier to their success.

First, I explained that **almost no undergrad has an exceptional foundation of the lab language at the beginning of their research experience**, so that wouldn't mean an extra disadvantage for them. Next, I reassured them that, if anything, they were already an expert in two languages, which was proof they would succeed in learning the lab language (which was going to be much, much less complicated than the two they already knew). Then I explained how the process of conducting research automatically becomes a language immersion experience and they would quickly learn enough terms to feel confident.

If learning the lab lingo is also one of your concerns, know that on the first day of your research experience, you'll begin to learn the name and definition of everything you need to know. Each item you turn off or turn on, or use to complete a procedure, will have a noun or verb associated with it (and sometimes both). Also, you'll learn some of the research language by learning definitions from the internet or a textbook, asking your labmates and in-lab mentor questions, and reading journal articles. This goes for every type of research experience in every discipline.

As for previous exposure, if you've taken lab courses in high school or college in topics similar to the type of research experience you choose, you're more prepared than you think you are because there will be some overlapping vocabulary. Plus, all STEMM foundation lab and lecture courses expose you to some scientific terms, which is helpful even if they all aren't the same ones you use during your research experience.

Even if you have no lab or research vocabulary to draw on, **because your language immersion experience starts on the first day of your research experience, it will get easier each week**, and eventually adding new words will become second nature instead of overwhelming. You might even start using some research terms in real life. One former undergrad remarked how they navigated on a road trip with friends who teased them for giving driving instructions such as "in zero point five miles, turn left" instead of "in half a mile, turn left." Another now refers to any time spent waiting for something in their life as an "incubation period."

Although it will take time, rest assured, you'll become comfortable using the technical language associated with your project. Until then, you'll experience a learning curve and might wrap up your research at the end of a day more mentally tired than you imagined you could be. **But not having a scientist's vocabulary right now won't stop you from finding a research experience or being successful in one after you do.**

I didn't participate in a research project in high school. How can I compete for positions in college against students who did?

There's no need to be concerned. Although some students participate in college-level research as a high school student, it's not the only form of experience that mentors consider when selecting undergrad researchers at the college level. Some mentors believe that specific lab or lecture classes (from high school or college) are enough experience and others have no prerequisites in this category. And there are even some mentors who hesitate to offer a research position to an undergrad who has already participated in several research experiences during their college experience on the same campus.

Also, it's important to understand that having experience—by any definition—isn't necessarily the golden ticket to secure an undergrad research experience. For many mentors, selecting a new team member is about finding someone who's genuinely interested in the research they do, demonstrates professionalism when they apply and interview, and has enough time available to commit to the experience. More details on what counts as research experience are covered later in this chapter in the section "The Experience Paradox."

I'm transferring from a community or state college to a four-year university. Can I be as successful in a research experience as a student who started their degree program at the four-year university?

Yes. Being a transfer student won't prevent you from achieving success as an undergrad researcher. One colleague, who directs a bridge program for students transferring from a college to a four-year university, remarked that **transfer students perform as well as—and sometimes outperform—undergrads who started a degree program at the four-year institution.** In addition, our colleague mentioned that at the end of a transfer student's degree program, their applications for grad, med, and preprofessional schools are as competitive as students who started their degree at the university. The same success holds true for transfer students who join undergrad research experiences. Also, remember that you were admitted to that four-year university on merit—you earned the opportunity to transfer.

What if my GPA isn't as high as my classmates'? Or what if it's low? Or what if I didn't do well in an introductory STEMM course? And if GPA doesn't matter, why do potential mentors ask for a transcript?

Many students believe, or have been informed, that an excellent academic record is the key to securing an undergrad research position. So, if a student has a much lower GPA than their classmates', or if they received a B or C in an introductory lecture or lab class, they may not choose to pursue undergrad research. And from posts on social media and some feedback from colleagues, undergrads in this situation who are also from underrepresented and underserved communities are more likely to be actively discouraged by academic advisors or others from pursuing an undergrad research project or a STEMM career than classmates who aren't members of these communities.

But as mentors with decades of mentoring experience, and who each failed a science course as an undergrad, we also know that **GPA or a failed class doesn't accurately reflect an individual's success as an undergrad researcher or their suitability for a career in STEMM.** And we know plenty of PIs (and grad students, postdocs, and staff scientists) who feel the same way. When we've asked other undergrad research mentors to weigh in on the minimum GPA required to join their group, responses have varied from "I never consider GPA" to a range of numbers. Granted, some mentors and some summer undergraduate research programs do place a high value on GPA, so there might be a subset of positions that you won't be eligible for if your GPA isn't in a particular range. And although our aim for this book is to get you into an *experiential* learning activity, some labs start students off as observers or as paid lab assistants. In these cases, a student who demonstrates a strong desire to learn, a growth mind-set, and reliability might have the option of moving into a research position even in labs that have a higher GPA preference. We cover these options later in this chapter in the section "Wet Lab Positions for Undergrads."

Still, most potential mentors will want to review a copy of your transcript as an opening to discuss a particular academic topic or for other reasons, such as to do the following:

- *Gain a generalized understanding of who you are as a student.* Have you completed courses that align with their research program? Are you interested in your major and will you likely be interested in the subject or discipline that is the core of their research program?
- *Ensure that you've completed prerequisite lab or lecture classes that are important foundations for conducting research with their group.*

And sometimes an interviewer might want to discuss a particular transcript for one or both of these reasons:

- *To evaluate whether you are overcommitted with activities.* That could be a concern if you have a history of abandoning commitments or trying to do too much at one time. (A potential mentor might not be concerned about a transcript hiccup if you work at a job or internship and don't have as much time for studying because of those responsibilities.)
- *To offer you help if you want it.* Although most interviewers won't want you to disclose personal information that you'd rather keep private, some will ask if there is anything you need help with managing if they think your transcript indicates you might need assistance. For example, if you share that you're struggling with classes, they might recommend tips to help you because many introductory STEMM courses require a different studying technique than typically used in high school courses.

However, it's also important to know that even a perfect GPA won't guarantee a research position or automatically make you more competitive than someone with a lower one. Through social media, and in office hours, we've connected with undergrads who were frustrated because they were unable to find a research mentor and couldn't understand why—after all, they all had a near-perfect GPA. But if your GPA were all that mattered, this entire book would be a single tweet along the lines of: "Tell all potential mentors you have a near-perfect GPA."

The take-home message is this: **don't disqualify yourself from a research position because of a less-than-stellar GPA or because you didn't earn an A in some STEMM classes, because plenty of mentors won't.** Also, participating in an in-depth research experience culminating with a strong recommendation letter can substantially support a future job or academic application by highlighting your achievements in other areas, as discussed in chapter 1.

I just switched majors, so I'm not an expert yet. Will this be a problem in a research experience?

Not at all. Even an expert can be temporarily stumped by something connected to their research project. Every scientist relies on several tools to contemplate the next steps in a research plan, interpret data, and learn unfamiliar techniques. Those tools might be the internet, journal articles, textbooks, mentors, labmates, colleagues from down the hall or across the world, and experts we connect with on social media. Not knowing all the answers isn't a problem—it's simply a challenge to overcome.

A classmate was required to write a research proposal before starting their project. What if I'm required to do it, too? I don't even know where I'd start.

Not all research experiences require a preproposal, but even if yours does, we have good news: it's highly unlikely you'll be on your own to complete this task. When writing a research preproposal is mandatory, typically the student collaborates with their future research mentor to get it done. In some cases, a student might also visit their campus Office of Undergraduate Research, Career Resource Center, or Writing Center to attend a workshop on how to write a preproposal or to obtain customized help. We're not suggesting that the process will be easy or without some challenges, but with a little help from someone with more experience, you'll be able to complete this task.

I'm premed and I've been advised to pretend to be pregrad or undecided because a lot of professors and grad students won't mentor premeds in research. Any advice?

We don't deny that some researchers prefer to mentor pregrad students or flat out don't accept premed ones. Sometimes the reason is that the mentor believes that their expertise and mentoring energy are best used to train those who plan to pursue a career in research. (However, we believe that the world needs good doctors, too, and if the training a student receives in our research group helps them get there, well, then we're proud to be a small part of it.) For other researchers, the hesitation might arise after working with a premed student who approached research as a checkmark to enhance their med school applications instead of as a genuine experiential learning opportunity. Some potential mentors have been advised by a colleague to avoid training a premed student because of the dreaded "premed reputation." The interpretation of the premed reputation varies but is sometimes translated as a student who doesn't value nonclinical research, or someone who will quit research if an A is in limbo in any class or withdraw immediately after their mentor submits a recommendation letter for med school. We understand not wanting to mentor a student who takes this approach to research, but in our experience these issues could arise from a student in any career path.

But the question is: Should you claim to be on the pregrad or the PhD-MD track to increase your chances of securing an undergrad research position? After all, research groups that don't train premed students rarely state so on their website or in an advertisement for an undergrad researcher. **The answer is simple: No. It's not worth it, or even necessary, to misrepresent your career path to find a research experience, and doing so could possibly cost you a research position or land you in one that you don't really want.**

First, understand that it's easy for an experienced research mentor to spot a premed student simply by reviewing a resume and transcript. For example, if in high school you volunteered at a hospital and participated in charity events to raise money for human disease research, and then in college joined the American Medical Student Association, shadowed in a clinic, and took a premed health professions class, you're obviously premed. However, if you balanced those activities with participating in high school science fairs and community education outreach, and then in college continued with science outreach and classes to prepare for grad school or similar activities, it's plausible that you could be premed, pregrad, or undecided. We do recognize that either pattern can be true for someone who isn't 100 percent clear about their career path.

Second, misrepresenting your career path can make your search more difficult because it could raise a character issue. Chances are, when you apply for a research position, the competition will be tough, so you don't want to raise any "red flags" at the application step—and misrepresenting your career path definitely raises one. Plus, the potential mentor who reviews your application will be less likely to forward it to a colleague who does mentor premeds if it seems as though you're misrepresenting something.

Third, misrepresenting your career path might land you in the wrong research position. In a research experience, both your happiness and your success are highly dependent on achieving your goals. We don't mean your research goals—although those are important. We mean the personal and professional development goals you want to achieve through an undergrad research experience.

Those goals should be achievable through the experience you join. As a premed student, you want your research mentor on your side—advising you throughout your experience on opportunities that will ultimately strengthen your med school applications. If you pretend to be on a career path that you aren't, the guidance from your mentor won't be customized to help you, and certain suggestions will be a waste of your time. Essentially, if your mentor doesn't know your true professional path and goals, they can't help you achieve them.

There are plenty of research groups that welcome premeds or don't care what an undergrad's career path is. Your goal is to find one with a research program that you're genuinely interested in joining and have enough available time in your schedule to uphold their required commitment.

THE IDEAL TIME TO START UNDERGRAD RESEARCH

No matter what type of research you want to do, or what your goals are, **start your research experience as soon as you can handle the time commitment without compromising your academics or well-being.** Some undergrads are able to start as a first-year student, but for many this additional activity makes their transition to college life overly complicated.

There will probably be several factors you'll consider when debating when the best time to start undergrad research is for you. Some influences will be personal, such as your academic load, whether you're working through college, or whether you have a chronic illness you must manage. Other factors will be influenced by the type of research you want to do (some projects are more time-consuming than others) and what, if any, the expectations are for doing research-related tasks such as writing a preproposal or constructing and presenting a poster at the end of the semester. There are other possibilities not mentioned here that might be relevant to your situation. Conducting undergrad research will be *an activity* in your college life, not your entire college life. You'll need to balance your academics, job, and life obligations and wellness as well.

What are the advantages of starting a research experience early in my undergrad career?

Typically, and up to a point, the more time and effort you put into a research experience, the more benefits you'll gain. (For details, review chapter 2.) In addition, some research fellowship or internship programs, such as many summer undergrad research programs (SURPs), accept only students who are early in their college career. Other programs require research experience *and* that a student be early in their college career. We won't sugarcoat it: these are highly competitive programs, and if you're interested in applying, familiarizing yourself with them early on is an advantage. (Details on SURPs, many of which are paid opportunities, are covered later in this chapter.)

Starting a search sooner rather than later is also an insurance policy of sorts. If you decide that your first research experience isn't the right one, or you want to participate in multiple types of research, you'll want time to explore different opportunities. In addition, it takes time to earn a recommendation letter from your research mentor that strongly supports your application to grad, medical, or professional school or be useful in a competitive job market. Waiting to start a research position until a month or two before you need a letter is risky—especially if you don't interact much with the PI before a job or other application is due.

Even if your ultimate goal is to write a senior thesis, know that more time in a research experience will give you more opportunities to learn and recover from mistakes without the pressure of a thesis deadline looming. More calendar time will also give you the opportunity to determine whether you actually want to commit to writing a thesis, which can be as much work as another college course. If you decide to write a thesis, the more familiar you are with the research project and the background and significance of your project, the easier it will be to get the writing done.

Therefore, start your research experience as early as possible in your college career, to get the most out of it and so you don't unintentionally make yourself ineligible for certain opportunities.

I've been told that upper-level students are preferred for undergrad research, so isn't it better to wait until I'm a junior to search for an opportunity?

Not necessarily. There are several reasons why a classmate might be unsuccessful in securing a research position, and academic level is only one possibility to consider. **Anyone, including a professor, who tells you that you must be at *X* academic year to start undergrad research is offering their *opinion*—not the hive's collective wisdom.** Unfortunately, when sharing their opinion, some PIs give the impression that they represent everyone when they are probably sharing the requirement for their research group or what they believe to be a universal truth because it mirrors their experience. Although do keep in mind that if they are referring to a *specific* research program, eligibility can be connected to academic level.

But don't put off the search for a research experience based on the belief that it's impossible to find a project as a first- or second-year student. Even though some mentors require upper-level students, many prefer undergrads who are early in their academic career. In some research fields, the learning curve is approximately the same for everyone, and the time (in hours and semesters) it takes for a student to go from being an assistant to a labmate to a self-reliant researcher working on an independent project is substantial. Because of this time issue, some mentors won't consider training a rising junior or senior. Essentially, some mentors might prefer first-year or upper-level students, and other mentors aren't picky. It's always mentor dependent.

So, if I'm a senior, is it too late to participate in a research experience?

No. It's not too late. Some mentors prefer or require rising seniors for research projects. However, depending on the type of research experience you want, it might be more of a challenge to find a research position and be able to finish a project so close to graduation. **If you're interested in a research experience that would require substantial training before you could work independently, the main challenge might be to find a mentor willing to choose you over someone who would be around longer.** From the start, you'll need to be direct and prepared to demonstrate that you're genuinely interested in becoming a valuable member of the research team.

On a personal level, the disadvantage of delaying a research experience—beyond not getting the full scope of potential benefits or the same opportunities to enhance your resume—is the pressure you might put on yourself to make a research position work. If you don't like the research experience from the start, you'll probably need to decide whether you can stick it out or should abandon the idea of having a research experience completely. If you start in a wet lab experience, it might be heavy on the observational side, with limited interactive work, or you might primarily assist others with research-related tasks. Still, these can be valuable opportunities to enhance your resume and learn about science as a process.

Although it's typically better to start early, don't give up on participating in a research experience even if you're a senior. Just prioritize your search as soon as you can. And don't let someone convince you that it's impossible to find a position. They might mean well, but they're probably wrong.

What if I haven't completed all the core coursework and lab classes? Aren't those necessary for me to understand a research project?

Sometimes. It's true that for some research positions a mentor will require prerequisite lecture or lab classes because they will help an undergrad researcher understand their project or bring foundational skills to their research experience. We also recognize that some mentors who work primarily in wet labs in the biomedical area might prefer students who have taken lab courses in chemistry, physics, biology, or other subjects because those students will have been introduced to common techniques and safety protocols. And some researchers who do certain types of field research have told us that they prefer (although don't require) students to have taken a course that teaches some basic field skills even if it's not exactly the ones they use in their program.

However, many mentors only require a genuine interest in the research, a determination to stick with a project through the tough spots, and a compatible research schedule—not specialized knowledge or skills to start. So, unless you read an advertisement for a research position that lists specific lecture, lab, or skills as prerequisites, presume that they aren't required for a particular position.

THE TIME COMMITMENT

Ultimately, if you want to be successful and earn a strong recommendation letter, the research opportunity you choose must be compatible with the time (and schedule) you have available for a research experience. When I (DGO) was an undergrad, the first lab I wanted to work in studied interferon signaling in human cells. But I was turned away because my class schedule and the research schedule of the grad student I'd work with were incompatible. Fortunately, my schedule was compatible with the grad student mentor in the next lab I applied to, and I thoroughly enjoyed the topic of genetics, so that opportunity worked out. **For most research projects, a mentor will have a nonnegotiable minimum time commitment in mind in weekly hours and semesters.** It's also possible that their availability to work with a new student, or the time period the research will be done (such as nights, weekends, or daybreak) will also be a factor. Before you accept a position, it's essential to determine whether the time requirement will work for you or you should pass on the opportunity because you're unable or unwilling to uphold it.

How many hours per week will I be required to commit to a research experience?

The time commitment could range anywhere from three to fifteen or more hours per week during the semester to full-time (around forty hours) during the summer months. **For research experiences that take place during a semester or quarter, the number of hours is often based on the average time it takes an undergrad researcher with similar skills or experience to make progress on a similar project.** However, paid fellowship and internship programs require a near full-time commitment for several weeks during the semester or summer months and might be less focused on achieving specific research objectives.

Overall, the weekly time commitment during the semester will be influenced by the project itself and the type of training needed to accomplish those objectives. If a project involves learning multiple techniques, for example, or even a single but particularly challenging one, then the time investment will be on the higher end. However, if a project requires minimal training, the required hours per week will likely be on the lower end. But there are other scenarios that could apply. If your research takes place at a field site, the commute time might be significant or inconsequential, which will have an impact on the hours required. And if you're registered to receive GPA credit for conducting research, there will be a certain number of hours required for each credit hour.

In addition, as mentioned previously, in some research experiences students are required to spend additional time constructing posters, reading scientific papers, planning experiments, and attending group meetings or research symposia.

How will my research schedule be determined? How flexible will it be?

Your research schedule is likely to change each semester along with your class schedule, project's objectives, extracurricular activities, and work schedule, if applicable. **The exact details of what your research schedule are impossible for us to predict and should be covered at an interview.** However, some possibilities are that you might be required to adhere to a specific schedule, with a set number of hours per week and per research session (whether done in a lab, clinic, or field), or you might have significant flexibility in managing your time. Perhaps your schedule will be inconsistent: long days intermixed with short days, but you won't know what kind of a day it will be until you get started. Maybe your schedule will have a significant unknown factor—you won't know if you'll have tasks to do until you arrive each day, or you might have the option of going in only when an aspect of your research project or an experiment needs tending. However, the following **two key factors will influence your research schedule:**

- *Your in-lab mentor's work schedule and your training needs.* If the majority of your training requires that your and your in-lab mentor's schedules overlap, then you'll likely need to fit your schedule around their availability. If the project requires minimal training or minor supervision, your research schedule might be quite flexible provided you're never working alone.
- *Your track record for getting things done and following safety protocols.* With time, and a proven track record of self-management and solid research skills, your in-lab mentor might agree to flexibility in your research schedule if you continue to meet project objectives and no safety issues arise. This flexibility could include working early mornings, at night or on weekends, or other scenarios but, again, provided you're not by yourself. For some undergrads this scheduling flexibility is essential to coordinate aspects of their personal life with a research experience. The number of hours required per week will likely remain consistent throughout your research experience, but they might increase if you choose to spend a full-time summer conducting research. However, some projects leave little room for schedule flexibility, regardless of your abilities or reliability, simply because of the types of procedures done or the location where the research takes place. Conducting surveys with medical patients, for example, might only be possible during clinic hours.

How many semesters will I be expected to participate in a research experience?

At the minimum, most mentors want a student to continue with research until specific project objectives or benchmarks are achieved. In essence, a mentor hopes that an undergrad will produce results or data or achieve certain project-related objectives that offset the time and productivity lost to train the student. For some mentors, that time is a single semester or a summer, and for others it's a year or more including a full-time summer commitment. Typically, a mentor will state the complete time commitment they expect at the interview or beforehand in a position advertisement. If they don't, you should ask.

What if I accomplish the project objectives and want to stay to do more research?

In research, as with most skillcentric activities, the more experience you have, the more success you tend to achieve. And because it can take both a long time and require significant effort to train an undergrad researcher, most PIs want to keep skilled, productive, and enthusiastic students around as long as possible. If a research project is ending soon, and you're happy with the experience, definitely ask your mentor if you can continue with a new project. But if your in-lab mentor is leaving the research team for a new adventure, connect directly with the PI.

However, some research experiences have a definite start and end date. For example, if a project involves collecting samples in the field during the summer, then the dates the work must be done are set. But if you're interested in the next part of the process, ask to participate in sample processing or data collection or analysis when the research moves into the wet or dry lab.

What if I want to make a shorter-term commitment than the PI or mentor requires (either in weekly hours or in semesters)? Should I accept the position anyway?

No. You should only apply to, or accept an offer for, a position if you believe you're likely to uphold the expected time commitment. Sometimes undergrads believe that if they continually express enthusiasm for a project, their mentor won't mind if they bail before completing the agreed-on time commitment. Other undergrads might have been coached by a well-intentioned but often misinformed advisor that the student can negotiate a reduced research schedule a few weeks after starting a research position by stating the need for more study time. Granted, upon reflection, most mentors would rather that a student move on from the research project than force themself to stay in a research experience that they don't like enough to produce high-quality results. And most mentors will understand if an undergrad needs to reduce their research hours to keep their academics strong or because a matter in their personal life needs to become their immediate priority. However, this doesn't mean that reducing the time commitment you made to the project will come without consequences, and in some cases, you'll risk a very real penalty. Therefore, honesty is the best policy when it comes to your research time commitment—for the ten reasons listed at the end of this section.

Does that mean I'm stuck if I accept a two-semester project and later I decide I don't like it, or I become overcommitted with my academics and need to reduce my hours?

No. Most mentors won't expect you to continue with a research experience if you detest it or if the time commitment threatens to compromise your academics, job, or mental or physical health. When an interviewer says something similar to, "I'd like you to make a one-year commitment," they generally mean, "I'd like you to make a one-year commitment *in good faith*." Your obligation is to accept the research position *only* if you're genuinely interested in it and if you believe that you'll uphold the time requirements barring unanticipated, insurmountable circumstances.

But you must not let the fear of joining the wrong project prevent you from pursuing a research experience. If you first choose a research experience that turns out to be incompatible with your goals or happiness, don't panic—you don't have to stay for years. Even if you registered for class credit and it's past the drop-and-add deadline, or you're part of a full-time summer research program, keep in mind that it's not for the rest of your life—it's at most until the end of the semester.

Also, remember that **research projects can sometimes be more frustrating in the beginning**—especially if you're in the process of acquiring skills or navigating information overload. **Several undergrads have told us that the first three to five weeks were the hardest in this aspect. So, if you can stick it out for a while longer, you might not feel like quitting.**

But if it turns out that the project or experience isn't a good match for you, know that there is value in trying something new, even if the main thing you learn is that you don't like it. Being able to cross items off your things-to-try list is as important as putting items on it. But fully use the search, application, and interview strategies in this book to increase the likelihood that you'll find the right research experience for you the first time around.

What if I don't know how long I want to participate in a research experience?

Right now, you don't need to know exactly how long you plan to participate in research, but it will make your search easier if you have a general idea. Fortunately, you've already established a foundation for determining this in chapter 1. When you start your search, you'll review the accomplishments you highlighted in that chapter to guide you.

TEN REASONS THAT HONESTY IS THE BEST POLICY WHEN IT COMES TO YOUR RESEARCH TIME COMMITMENT

1. **It's harder to care about a project if you've checked out before you've even started.** If you accept a research position you don't really want with the intention of leaving when something better comes along, or with the idea that you'll quit as soon as you can list the experience on your resume, it's unlikely that you'll be happy even in the short term. If you're only at the search or interview stage and you're trying to think of ways to avoid the required time commitment, you haven't found the right research position and need to keep searching.
2. **The hours might not be negotiable.** In some research opportunities, if an undergrad can't uphold the agreed-on time commitment, there may not be an option to continue with the project. This might be because a mentor recognizes the importance of maintaining an academic and life balance and they want to prevent a student from getting behind in classes or becoming stressed out trying to manage too many commitments. Sometimes, however, it's because a certain number of research hours per week are needed to gain technical skills and make progress, and thus reducing that number will set the student up for failure. Therefore, when a mentor requires "*X* hours per week," they might be fine with regular time off or reducing your weekly hours—or they might not. But you should presume that it's not too flexible. Use the strategy presented in chapter 4, in the section "Step 1: Schedule and Prioritize Your Time," to help determine how much time you actually have for a research position.
3. **The student contract is attached to a specific number of hours pers week.** If you are taking research for GPA credit, the research probably requires a set number of hours to earn that credit. Skipping too many hours may negatively impact your final grade.
4. **The training plan is often designed with a specific time commitment in mind.** If a project is to be completed over multiple semesters, there could be several distinct training aspects to it. The first semester might only include observing others, reading journal articles, or writing a research proposal. Alternatively, a project might be all benchwork the first semester, but only to learn skills or build the tools that you will use in experiments during the second semester.
5. **You'll miss out on the interesting stuff.** Some mentors wait to put a new undergrad on an independent project or something "interesting" until the student has demonstrated genuine enthusiasm, reliability, and a dedication to research. Although lab volunteers don't have timecards, if you miss enough hours, you might never get past the initial test phase to the interesting research phase. (The number of hours missed before it becomes an issue with the PI or an in-lab mentor varies.)

6. **You'll risk being thought of as unreliable by your labmates.** If you show up at the lab sporadically, or for less time than the commitment you agreed to at the interview, it's possible that other group members will need to cover your responsibilities. If you continually show up late, leave early, or are altogether absent without connecting with your in-lab mentor about how to resolve this issue, eventually some of your responsibilities (or a part of your research project) might be permanently reassigned to another researcher. Once this starts, it's easy to lose enthusiasm for conducting research, and the results might initiate a vicious circle—you'll have fewer tasks to do, so you reduce your research hours even more, which leads to even fewer responsibilities. And note that it's extremely difficult to regain your labmates' trust once you've been unreliable, and that can happen quickly—especially if their productivity is dependent on yours or you routinely volunteer to do tasks and then fail to follow through on completing them.
7. **You'll potentially sacrifice a recommendation letter.** Strong recommendation letters aren't guaranteed because you join a research group. Some PIs don't write letters for students who aren't around long enough (in hours or semesters) to demonstrate much professional or interpersonal development. And if they suspect that your only goal of joining a research group was to secure a letter and that you had no intention of upholding the commitment you agreed to, a reference letter might not be on the table. So, it's a risk to take a position knowing that you won't uphold the commitment—especially if your main goal is to secure a letter at the end of a single semester.
8. **You'll lose an opportunity of a potential authorship.** If one of the project's objectives is to produce specific results or data for a publication, and you end the research experience before you've made much progress, you could lose the chance of authorship on the resulting publication. Authorships don't happen because you worked on a specific project for a certain amount of time or because you're a member of the research group; authors must make a meaningful contribution to the study.
9. **You might lose a scholarship or fellowship.** Some awards aren't distributed until certain requirements are met, and other awards can be rescinded. Award requirements might range from submitting a final report or paper, to attending a scientific conference, to putting in a preset number of hours on your research project during a specific period. For some awards, failure to attain the requirements will automatically disqualify you from receiving the official award or the full benefits of it.
10. **It's unfair to your mentor.** By planning an undergrad research project, your mentor makes a commitment to you before you even start on a research project. As soon as you officially get started in your research experience, their efforts continue. It's rude to accept a position if you have no

intention of upholding the commitment. We cover more on this topic in the final section of this chapter, "Why Research Positions Are Competitive (and What You Can or Can't Do about It)."

THE EXPERIENCE PARADOX

Nothing is more frustrating than to be told that you need experience to get a research position when you can't get experience without having held one! Fortunately, *experience* is a broad term; there are research positions that don't require it, and some mentors even prefer undergrads without it.

How much research experience will I need to get a research position?

It depends and varies greatly among mentors. **Some mentors require a specific skill set or research experience to be considered for a position.** Experience in those cases could be a specific lab class, working knowledge of a particular technique, or time spent in a professional research lab either in college or as a high school intern. Even in the same research group, there could be variation. One mentor might require experience with writing code, whereas **another mentor might prefer to teach a new undergrad everything they need to know.**

So where does that leave you? Essentially, it's all up to the mentor. So, for some positions, yes, you'll need specific experience to be considered, but for other positions, an untrained student who demonstrates a genuine interest in the research project and has the right schedule will fulfill all the requirements. **In many disciplines, most undergrads start their research experience with only lab class experience.**

What is it like to start a research project without research experience?

Surprisingly, it's similar to starting *with* previous research experience. Everyone feels awkward in the beginning. It's a new environment, with new procedures, new expectations, and new people. However, if you have little to no experience, you won't be expected to be an expert in techniques or know how to keep a proper field journal or lab notebook. You'll probably find labmates willing to guide you in basic protocols as long as you are reliable, follow instructions, and show dedication to learning what they teach. **As with any new situation, you'll adjust, and it**

will get easier day by day. You'll also generally find more than one person who will make you feel welcome as you start to feel like a member of the team.

How difficult will it be to learn the research techniques?

Unfortunately, there is no easy test to determine how hard it will be to learn the techniques and gain the technical skill set needed for a given project. Research, depending on the discipline, can have a steep learning curve—more so for some people than for others. We can reassure you, though, that all undergrads we've worked with have learned the fundamental research techniques in our lab if they were willing to put in the effort and follow instructions. Most researchers who mentor undergrads use a supportive teaching approach to guide new students through the learning process. However, if your in-lab mentor isn't supportive (and other group members aren't a source of help, either), we recommend searching for a new research experience where the training culture *is* encouraging. We don't recommend staying in a research position where there isn't mentorship geared toward accomplishing your goals or one that makes you feel as though experiencing failure means you're not cut out for conducting research. (Remember, *everyone* experiences failure in research.) However, if you have a supportive and engaged in-lab mentor, even if you struggle to get a technique to work, you'll eventually overcome the issue and be successful. Yet even with a supportive mentor, **overall, it's likely that your progress will depend on the following:**

- Your dedication, self-motivation, and perseverance in acquiring the fundamental skills needed to do your project
- Your desire to understand and follow instructions, and the self-reliance to carry them out with minimal supervision or assistance
- Your patience, self-discipline, and perseverance when conducting repetitive work, if applicable to your project
- The time commitment you make to your research experience and how you use your time when working on your research project
- Your determination to overcome fear and ask questions when you're unsure of how to do a procedure or interpret results and to take detailed notes when your labmates provide answers—notes you absolutely must be willing to refer to later to answer many of your own questions rather than ask your previous questions several times
- Your level of desire to participate in the research experience and do the tasks associated with it

What if I make mistakes?

One of the few guarantees about research is this: you will make mistakes. After all, you'll be learning new procedures, acquiring new skills, and discovering the boundaries of your knowledge. Making mistakes is simply an integral, unavoidable part of this learning process. **However, the biggest mistake you could make is to let fear stop you from pursuing a research experience.**

As you gain experience, your mistakes will decrease, and the success you celebrate will increase. On your first day, you'll know the least amount of information and have the least number of skills. Each day after, you'll know more and be a little more confident than the day before. Then one day you'll have an incredible realization along the lines of, "I got this," and it will be your best research day ever.

USING LAB CLASSES TO PREPARE FOR UNDERGRAD RESEARCH

In every lab class, you're introduced to a type of research. But there is considerable variability among lab classes—different disciplines, different techniques, and different emphases. Lab classes vary from those that focus on general or introductory techniques of a specific discipline, to those that integrate techniques from several disciplines (such as biology, chemistry, and physics), to those that focus on a single research objective that allows the students to conduct research and cowrite a manuscript and submit it for publication at the end of the term. Some lab classes emphasize the scientific method, the importance taking accurate and detailed notes, experimental design, or training in field skills.

There are advantages to completing lab classes prior to starting an undergrad research experience no matter what your project turns out to be. In addition to being exposed to the aforementioned topics or activities, a lab class might help you acquire familiarity with certain research equipment or glassware, tools, or analytical programs; safety procedures; or the technical language (sometimes referred to as *jargon*) associated with the discipline. **A lab class also can help you decide whether you like (or dislike) a set of techniques or a type of science.** This knowledge will be invaluable when you start your search for a research project. Over the years, several undergrads have joined our lab after finding inspiration from the techniques they learned in a general biology, microbiology, genetics, or other lab course. Conversely, we remember one student who remarked that the techniques they learned in a biology lab were "a tedious waste of time," so we knew that they wouldn't be happy doing a project with our group.

Will lab classes teach me the techniques I need to conduct research?

Maybe. It will depend on what techniques you learn in the lab classes and the research project you join. If you took a chemistry lab and your research project is about wildlife ecology and conservation, you'll probably find that the techniques you learned in the chemistry lab, although still valuable, aren't directly applicable to the project. But if you took a lab class that focused on sequencing the genome of an organism, then you would be prepared for a variety of research projects that incorporated similar techniques. Or if you worked with bacterial stocks in a microbiology lab and your research project involves bacterial manipulation, your instructional lab experience could be directly applicable.

If I learned a technique in a lab class, will I be able to do it by myself for my research project?

Perhaps. It mostly depends on the type of lab class you took and how much of the technique you completed in the class yourself versus how much the TA or professor did for you. You may not realize it, but professors and TAs work hard behind the scenes to make sure lab class sessions go as smoothly as possible. In lab courses that could be mostly categorized as a wet lab, professors and TAs might do several steps of a technique or experiment for you before you arrive or after a lab session is over. Therefore, **you probably won't be an expert after learning a technique in a lab class, but you might have a substantial introduction to benchwork related to your project.** It's similar to studying a language that is foreign to you before conversing with someone who grew up using it. You quickly realize that conjugating verbs and stringing words together to have a meaningful conversation is more difficult than anticipated but much easier than if you had not taken previous language classes. This is important to keep in mind so that you'll be patient with yourself if the first few weeks in a research experience are a little bumpy. You don't want to give up if it turns out that you aren't as prepared as you had believed you were. Knowing the name of a technique and being familiar with it isn't the same as knowing how to execute it from start to finish with a protocol that is slightly different from one you used in a lab class.

In the lab class I took, the experiments failed sometimes, so I didn't learn how to do a technique or an experiment that was listed on the syllabus. Will that be a problem if the PI only accepts students who have completed the lab course?

All PIs know that sometimes research procedures fail—even in a lab class. And most PIs don't expect an undergrad to be an expert after taking a lab class as much as to have the basic knowledge or technical proficiency the class provides. And for some research experiences, there is an unexpected bonus if some techniques or experiments failed in your class. As odd as it seems, to have firsthand knowledge that even the most well-planned experiment or technique can go awry is an advantage. **Undergrads who start a research experience with this understanding are more resilient (and often more patient) when faced with failures in their research,** and that makes the transition to a professional research lab easier.

How do lab classes and professional research labs differ?

It's impossible to characterize all the potential differences among all lab classes and all research labs. Plus, as you know, each research group is a unique entity with a unique culture, and their research might be conducted in the field, or in a dry or wet lab, or a combination of those. However, there are some general differences among lab classes and professional research labs that often catch new undergrads off guard. Although several of these are closely related to research done in wet lab environments, many are applicable to all research experiences regardless of the space where it's conducted.

NINE POTENTIAL DIFFERENCES BETWEEN THE UNDERGRAD EXPERIENCE IN A LAB CLASS AND THE UNDERGRAD EXPERIENCE IN A PROFESSIONAL RESEARCH LAB

1. **Scheduling lab time.** Out of necessity, most lab classes have a start and stop time for each session. If your lab class is scheduled for noon to 3:00 p.m., you're out the door by 3:00 p.m. so the next class can begin its session. In wet lab research, however, your schedule is influenced by the procedure you do each session. For example, if you're scheduled from noon to 3:00 p.m., and your experiment isn't finished by 3:00 p.m., for whatever reason, you'll generally stay until it's at the correct stopping point,

regardless of what the clock says. In some instances, a labmate might be able to finish up for you, but this shouldn't be your go-to solution if you want to develop an independent project and earn the strongest possible recommendation letter. Quitting an experiment early can be expensive in time and lab resources, especially if a labmate's experiment is based on yours. If you end up consistently having issues with leaving the lab on time, you'll need to connect with your in-lab mentor to find the best solution. However, if you're conducting research at a field site, especially if the research is seasonal, the amount of sunlight (or moonlight) or the distance you're required to trek to do your research might influence how long your research day is.

2. **Moving forward.** In a lab class, you might wrap up a set of experiments in a session or two, and you might never do the same technique twice. In a research experience, you might use the same set of techniques for several days or weeks (or longer) before moving on to something new, or you might learn a core set of techniques that you recycle throughout the experience.
3. **Preparing supplies and equipment.** Locating what you need is easy in a lab class, where reagents and supplies are often at your research bench or close by when you arrive, incubators are set to correct temperatures, and stocks are often ready to dilute or use. In a research lab, you'll need to learn where the community supplies and reagents are kept, retrieve them when needed, and put them back when done. You'll likely learn to prepare many of the solutions you need, starting with the solid chemical, a scale, and the correct type of water. In addition, you'll likely be responsible for programming equipment and possibly maintaining personal stocks of the organism you use, such as cell lines or seed, worm, bacterial, or fly stocks. And you'll need to do all this while coordinating equipment use and bench and freezer space with all your labmates who are also trying to complete their research objectives.
4. **Designing experiments.** In a lab class, procedures are designed with careful thought given to the amount of time available in a session and the presumed technical skill level of the students. In a research lab, although a student's level of expertise is sometimes considered when assigning a project, experiments and procedures are selected to achieve project objectives, and some techniques can be particularly challenging until certain skills are acquired.
5. **Keeping notebooks.** In a lab class, your notebook is turned in for grading, but often it's yours to keep at the end of the semester. In addition, you'll receive instructions from the lab manual or the TA as to what information needs to be recorded. **In a research lab, "your" notebook doesn't belong to you—it belongs to the college or university or to your PI, and**

some PIs have a strict policy that it never leaves the wet or dry lab. Your research notebook is as essential to your in-lab mentor and the PI as the results and data you produce. In addition, the information you need to record in your research notebook isn't always immediately obvious. Often guidelines, rather than specific instructions, are offered, and confusion as to what exactly needs recording accompanies the start of each new research project.

6. **Working outside the lab.** In a lab class, you'll likely be required to read a protocol or an experimental overview before class, write lab reports, and possibly prepare a poster or do an additional out-of-class assignment. In a research lab, you might be expected to read some of the PI's papers, learn background material about techniques or your experiments, create and present a poster at a meeting or symposium, learn about your labmates' projects, or design an experiment.
7. **Knowing the plan.** In a lab class, the syllabus and lab manual make it easy to learn the objectives of each lab session before arriving. In a research lab, your activities for a particular day might be unpredictable. When you arrive for the day, your cultures might not be ready for processing; your cell lines might be contaminated; or your mentor might have you work on an unexpected task. Your research strategy might even change in the middle of the lab session because with a research project, being prepared also means being prepared to change your plans when needed.
8. **Obtaining help.** In a lab class, you can rely on a professor or TA (or two) to quickly help when you get stuck or need clarification. In a research experience, your mentor might not be available to help you throughout your lab session, or their time in the lab might not even overlap with yours. You might need to postpone a procedure or set up an appointment with your mentor outside your research hours if you get stuck.
9. **Repeating experiments.** If an experiment or technique doesn't work in a lab class, the TA might instruct you to borrow supplies or notes from another student or group, or they might tell you what the results should have been and direct you to move on to the next module. In a research lab, you'll likely need to start the procedure over if you're unsuccessful the first time. Or the second time. Or the third. Basically, it might take multiple attempts over several weeks to achieve the desired result before you can move on to the next step.

WET LAB POSITIONS FOR UNDERGRADS

Research groups that work in wet labs (either as their primary space or in addition to a dry lab or field site) don't always start new undergrads on a research project. Some research groups only bring in new undergrads as observers or general lab assistants. Sometimes a mentor will offer an undergrad the option of selecting which opportunity they prefer. In this section, we discuss each position and the general responsibilities associated with them. Although some of the information presented next is applicable to a student in a dry, clinical, or field research experience, it's more common for students joining projects in those categories to start as a researcher from day one even if their primary responsibilities at first include reading journal articles or receiving technical training.

Broadly considered, there are three potential roles available to an undergrad interested in joining a wet lab: an observer, a researcher, or a general lab assistant. Within these categories, a student could be a volunteer, a student registered for GPA credit, or, in the cases of researcher or general lab assistant, an employee. The responsibilities associated with these positions aren't necessarily distinct as there can be substantial overlap among them. If you join as a volunteer, for example, you might spend most of your time observing your labmates as they work but also be assigned some research responsibilities. Or you could be a paid researcher but still do a lot of research-related chores because your experiments produce mountains of dirty labware. And if you're brand-new to some types of research, your training might involve observing labmates while you learn how to properly carry out certain techniques even though you're officially a researcher.

No matter which position is yours, from the start of your experience, show up on time, ready to work (or observe) and to learn. If you're given an opportunity to contribute in some way, follow your labmates' instructions and ask questions for clarification, if needed. **Your overall goals are to demonstrate that you have time to be in the lab, you want to be there, and you appreciate your labmates' ongoing efforts to teach or mentor you.** In each section covered next, we'll cover how to do that.

OBSERVER

If you accept this position, you'll mostly observe your labmates as they work, ask questions when it's convenient for them, and, we hope, be offered opportunities to do some research tasks. Also, you might be given some scientific papers to read that were published by the research group or by other groups. **Your goals are to learn everything you can, determine whether you'd like to remain a member of the research group, and demonstrate an interest in the research program and what your labmates offer to teach you.** When setting your schedule, approach it as seriously as you would if you were starting as a researcher. With this strategy, your labmates will be more willing to bring you into their research bubble and involve you in the inner workings of the research group. During this time, they will determine whether you're likely to be a reliable member of the research team, how easy you are to work with, and whether you're genuinely interested in what they have to offer.

RESEARCHER

As a researcher, also frequently referred to as an *undergrad research assistant*, most days will probably involve a combination of conducting benchwork, updating your notebook or taking other notes, and contributing to the overall operation of the lab by completing research-related tasks. (For additional possibilities, review the chapter 2 section "Nine Parts to a Research Project, Simplified.") You may also spend some time of your downtime observing your labmates as they perform a technique or assisting them on tasks. You might work on an independent project, or you might primarily assist a labmate who has designed a subproject of their main project or help whoever has a task that needs to be done that day. In some research positions, before the benchwork begins, there are training requirements—perhaps reading journal articles or learning a computer program or doing repetitive "skill drills" with certain wet bench techniques.

GENERAL LAB ASSISTANT

This position is sometimes referred to as a *dishwasher* or *lab assistant* position. Students who accept this position are sometimes hired as an employee and assigned the title of undergrad technician. Often, a general lab assistant position is a paid one, and primary responsibilities are centered on completing tasks that help keep the lab functioning, such as washing labware, making media, or performing other lab-keeping tasks. **If you plan to use this position as a launchpad to a research position, confirm with person who interviews you that it's possible *before* you accept the position.** Not all general lab assistant positions include the opportunity to incorporate a future research project.

In addition, you'll need to know how much the job pays and if the amount you're eligible to earn in a semester or academic year will be capped. Also, be aware that some labs offer a general lab assistant position as an unpaid one, so ask for clarification if you're unsure which type you're applying to. (In field research, this can also be a paid or volunteer position under the title of field technician.) It's reasonable to expect payment for this type of work and to take a pass on a research group if a paycheck isn't an option. Specific questions to ask about this position at an interview are provided in chapter 6.

Why would I accept an offer to join a lab as an observer instead of a researcher?

Don't underestimate the significance of an observe-to-learn opportunity. This is a common strategy that some PIs in wet labs use to screen potential researchers, while introducing students to the inner workings of the lab. It allows both the student and their potential in-lab mentor to determine whether they like working with each other, and for the student to decide whether the research is truly something they would like to do and if they like the lab culture. Understand that a PI will only offer this opportunity if they believe you'll be an asset to the research team—not because they are being nice.

For you, this opportunity can be substantially beneficial because as soon as you're in the lab, even for a few minutes, you've learned something. You've learned a little about the lab's research focus, how experiments are carried out, details about specific techniques, and most details about the lab culture. If you have the opportunity to help with an experiment, you'll gain some experience related to what you'll do if you stay in the lab as a researcher. Most important, **you'll have the opportunity to evaluate whether you're in the right research experience before you make a formal commitment.** Even if you decline to continue in the lab after this probationary period, the opportunity to observe, learn, or gain experiential learning won't be wasted. Experience, knowledge, and training are benefits that become advantages—even if they are from a short-term experience.

One additional note about joining the lab as an observer: Students from low-income backgrounds or those working to put themselves through college may not be able to accept this type of opportunity. If this describes your situation, it's worth asking the PI if you could register for GPA credit to conduct research, or if there might be a paid general lab assistant position or paid research opportunity, which are covered next.

Which position should I pursue if I want a paid lab position?

Although earning a paycheck for conducting undergrad research is a bonus for some students, for other undergrads it's a necessity. In particular, students from low-income backgrounds and students who work to put themselves through college (or to support their family) often can't afford to do research unless offered a paid position. Some PIs hire undergrad researchers, and others pay only students who work in general lab assistant positions. It's safe to bet that no PI will hire someone as an observer because most PIs are prohibited from doing so by their institution's or funding agency's policies.

Summer research programs that pay undergrads a stipend to participate and general lab assistant positions are more common avenues for earning a paycheck. However, if you have a financial aid award that can be used to supplement a portion of your wages (such as a work-study award in the United States), this might make it easier to secure a paid research position. (Check with the financial aid office on campus for your options.)

Also, some colleges or departments host programs that pay students a modest stipend to conduct research during the semester if the work is done with a program-eligible research group. These programs might be restricted to a specific department or one that includes multiple research groups throughout the campus. Unfortunately, these programs aren't common, so this possibility might not be available to many of our readers. Another option available to some students is a SURP, which we cover later in this chapter.

Why would a PI hire a general lab assistant? Isn't getting research done the most important thing? Wouldn't it be better to hire only researchers?

To the PI, the advantage of hiring a general lab assistant is this: **Every wet lab has a set of chores no one likes to do but which are essential to keep the lab functioning smoothly.** Dishes need to be washed, glassware needs to be sterilized, and numerous other research-related tasks must be done. Even in labs with access to an automatic dishwasher, someone must load the glassware, start the cycle, and put the clean dishes back in the cabinets. No matter how efficient all the group members are with these tasks, there is never enough time to get them all done. Hiring a general lab assistant to do some of those chores benefits the entire lab as soon as the person starts.

What is it like to join a lab as a general lab assistant?

At the start, you'll most likely wash a lot of labware. If you do that well, you might learn a few additional research-related tasks such as how to make media or prepare stock solutions. Then you'll probably wash more labware. After you demonstrate reliability and consistency, you might learn a research technique either by observing someone or assisting them. But if you're in a general lab assistant position, the majority of your responsibilities will be those discussed at the interview.

How can I go from a general lab assistant to a researcher position?

If the general lab assistant position can be used as a launchpad to a research position, then what you'll need to do to make that happen is simple. **Do everything you're instructed to do and do it well. Do it well every time. Every time. Every. Time.** No matter how boring the task. If moving into a research position is possible, then your performance as a general lab assistant will be a test. It will be a test of how well you follow instructions and follow through with tasks, your ability to work with others, and your self-reliance. Every research project has at least one boring phase and often several boring phases. If you demonstrate to the PI that you'll do a high-quality job on the most boring task, even after you've done it a thousand times, they will know that you won't slack when you hit a boring or repetitive part of a research project. Essentially, they will know that you're a good candidate for a research position.

However, if you complain about the work, or you show up late or irregularly, or you do a shoddy job, or you constantly ask labmates if you could do something other than what you were hired to do, then you'll prove to the PI that they made the right decision not to put you on a research project. In addition, it won't be long before you're dismissed from the lab because PIs tend to become annoyed when the simple things aren't done well or when they pay for subpar work—especially if your carelessness in cleaning the labware leads to contamination of a sample and forces others to waste their time and the lab's supplies in redoing the research. **Let us assure you: a PI won't move you onto a research project** if you don't meet their expectations in a general lab assistant position, or **if you show your labmates that you resent doing the work you agreed to do when you were hired.**

Over the years, we've hired several undergrads as general lab assistants. In most cases, the student excelled and remained a member of the lab as a researcher. If you deliver a solid, consistent performance as a general lab assistant with the understanding that becoming a researcher is a possibility, it will probably happen within the first semester, or at the start of the second. But you shouldn't be left guessing about when this could happen—ask specifically about the timetable at the interview.

What if I'm offered a choice? Which position should I choose?

If you're given a choice at the interview, it's important to consider your goals and life circumstances before making the decision. You'll need to evaluate whether the offer will give you the opportunity to learn without compromising your values, academics, or financial aid package.

However, we recommend that you don't stay in an observational or an unpaid general lab assistant position for more than one semester—at most. In general, it's reasonable to expect to be involved in regular, authentic research in a lab after a few months. So, if it's been that long, and all you've done in an unpaid lab position is wash labware, perform other random lab-keeping chores, or observe labmates as they work, it's time to ask for an official research position that is either paid or for course credit. Granted, washing labware and doing other research-related chores are endless for most researchers—students and professionals alike. But **a research experience needs to include actual research if you're going to gain the interpersonal and professional development from participating.** If your request to work on a research project can't be accommodated shortly after you ask, it's probably time to search for a different opportunity.

PROJECTS FOR UNDERGRAD RESEARCHERS

The types of projects available to undergrads in STEMM are as broad and varied as there are scientific disciplines and research groups.

In one research group, the team might have multiple ongoing projects, each of which may have one or more subprojects that support the lab's research focus. For example, a lab's overall objective might be to understand how microtubule organization controls plant cell wall expansion. One project might use in vivo localization of the microtubule severing protein, katanin, during epidermal cell expansion. A parallel project might be the analysis of mutations that enhance a specific, well-studied katanin mutation. A subproject of that might be the map-based cloning of an enhancer mutation, and another might be the transformation of an enhancer mutant with a green fluorescent protein–tagged katanin fusion. Each of these projects and subprojects has the ultimate goal of providing new knowledge about how microtubule organization controls plant cell wall expansion.

In another research group, there might be more of a unifying theme or tangential connections among researchers' projects, but not all projects support a single, big research picture. So, several group members might work on projects that are wholly independent of each other but have commonalities in the knowledge they produce.

As an undergrad, you might participate in multiple projects, observe, or assist on a single project, or you might be assigned your own subproject as independent research. Within this framework, what you work on might change during your research experience.

What if I want to ask my own question and design my own project?

Answering your own question, or investigating a problem that intrigues you, can be an incredibly rewarding experience. It's exciting to be an undergrad "PI" with the challenges and responsibilities that accompany this role. **You might find a PI willing to sponsor your research from the start, but it's also possible that you'll be required to first volunteer in their research group for a while.** If you receive the go-ahead to design your own project right away, you'll probably discuss your research ideas with the PI, do background research (in the library or online), and write a proposal that outlines your project objectives and the techniques you'll use to accomplish them. In some instances, the PI will be listed as your official mentor on record, but your project might still need approval by an administrator such as a director of undergrad research in your department. Although you'll work under the guidance of a PI, choosing this route will take a special amount of dedication. Therefore, it's important to thoroughly consider whether you're prepared for the level of independence and self-reliance that this approach requires before you start down this path.

Do all research opportunities include an option to ask my own question?

No. In some research experiences, there is no chance of designing your own project—even if you start with the research group as a first-year student and stay until graduation. Even grad students and postdocs don't always choose all their own research questions, even if they ultimately design and conduct the studies to answer them. However, many research experiences include a hybrid project that allows a significant amount of self-reliance and self-directed work.

What is a hybrid project?

Many PIs use the term *independent project* for research experiences that fit in this category. **It's a balance between having complete control over your own project and being an assistant on someone else's project.** It's likely that when you acquire certain skills, demonstrate self-reliance, and achieve your initial benchmarks, you'll be given reasonable autonomy with respect to which research strategies to use in your daily research, or possibly which subproject to work on. You might also have input as to the direction your project will take. In a sense, this is answering your own question—it's just within the framework of the lab's research focus.

If I don't want to design my own project, will it be more difficult to find a research position?

It depends. Working on a project designed by the PI (or an in-lab mentor) or on a project within the hybrid project category is common—especially in biomedical labs. In many of these kinds of labs, there are several ongoing projects that could be matched to a student depending on the research group's current goals, sometimes the student's current technical or analytical skill set, and often the amount of time the student can dedicate to research. (Remember, some projects have a nonnegotiable time commitment.) However, there are some PIs who require undergrads to contact them with an idea for a research project based on the current work being done by the group. In these cases, it's particularly important to spend time learning about the PI's faculty interests and crafting an Impact Statement for inquiring about a research opportunity. (You'll learn how to find out about faculty interests in chapter 4, and how to write an effective Impact Statement in chapter 5.)

How are projects prioritized?

Ultimately, in many research groups, the PI sets the lab's research priorities, and those priorities can change depending on a variety of factors. Even within the course of a few months, one project may become irrelevant while another becomes the new top focus. New data, a disproven hypothesis, or impending grant or paper submissions could influence why some projects are set aside while others are given higher priority. Even project objectives can change. Because research is about discovery, new results and data often produce a new set of questions to be answered. With that come new directions, possibilities, and decisions to be made. On occasion, a project may even be abandoned. However, we do need to mention that if a grad student or postdoc has secured their own research funding to pursue a project, they might do that work independently of the PI's priorities.

Will the type of project dictate whether I work as part of a team or on my own?

You'll always work as part of a team whether you design your own project or work on one created by the PI or an in-lab mentor. However, your project might or might not be an independent study. This is both lab-dependent and project-dependent. In some research groups, a single project is distributed among several researchers, and each person works on a designated part. Within that framework, each part could be independent of the next or could be intertwined so that part A is completed by one researcher and then part B is completed by another. Other possibilities include a project with significant collaboration with a specific member of the research team (such as your in-lab mentor) or a project that requires you to work mostly independently with occasional consultations from an experienced member of the group or the PI.

What if I want to work on a different project after being in the lab for a while?

How easy it will be to change projects, or if it will be possible at all, will depend on the reasons you want to switch and other circumstances. Although most research groups have multiple ongoing projects (and many grad students and professional researchers work on more than one at a time), that doesn't mean the PI or your in-lab mentor will automatically support your desire for a change, or that someone will be available to train you on something new. And if you're participating in a summer undergrad research program, which is covered later in this chapter, a new project probably won't be possible.

DECIDING WHETHER TO REGISTER FOR COURSE CREDIT FOR RESEARCH

Whether you register for course credit for conducting research is only partially up to you. Not all departments offer course credit for undergrad research, and some PIs don't permit a student to register for credit until that student has demonstrated a dedication to the research experience or a particular skill set. However, other PIs or departments require students to register for credit starting with their first semester of undergrad research. In addition, if you're to write a senior thesis, it's reasonable to presume that registering for class credit will be required for at least one or two semesters.

Before registering for course credit, it's important to have a solid understanding of the requirements of the position and how registering will affect you academically and possibly financially. Therefore, you'll want to consider several factors and consult with your academic advisor before making the decision.

START WITH THE BASICS AND CONSIDER YOUR ACADEMIC GOALS

Make an appointment with your academic advisor or the undergrad research coordinator in your department for specific advice on the benefits of receiving class credit for doing research and to learn what all your options are.

You'll need to know whether registering for undergrad research will count toward your electives, and how many credits (if any) will count toward your major and degree. You'll also need to know whether your research grade will factor into your GPA. Sometimes undergrad research is graded on the A, B, C scale; other times it may be taken as pass/fail or satisfactory/unsatisfactory.

Also, if it's offered by your department, inquire about registering for a 0-credit research option. These opportunities allow a student to list research experience on their transcript but don't require the payment of tuition or fees. However, be aware that the 0-credit option won't factor into your GPA or count toward major requirements.

With your academic advisor's input, put together a long-term plan that includes how many research credits you can take in total and when it would be most appropriate or advantageous in your undergrad career to register for GPA credit for research. But be prepared to be flexible. After reading the next sections, you might realize that waiting to take research for credit is the better option for you.

LEARN THE DEPARTMENT'S REQUIREMENTS

Most departments that offer course credit for undergrad research have a learning contract that covers the official requirements. **Typically, there will be a set number of hours a student is required to spend in the lab, in the field, or working on their undergrad research project per registered credit hour. There also may be additional requirements.** For instance, some departments require students to write a research proposal before starting their project, or a paper that summarizes their research results at the end of the semester, or both. If the department hosts a year-end research symposium, creating and presenting a poster or giving a short talk might also be required.

LEARN THE PI'S REQUIREMENTS

Sometimes PIs have additional requirements not covered in the departmental research contract. Although it varies, it wouldn't be unreasonable if you were expected to turn in your notebook or other research notes, or make a backup copy; clean your research spaces; and properly label, organize, and store all research samples prior to a grade being assigned. These are likely expectations regardless of your registration status. Some PIs also require a minimum number of participation hours per week regardless of the number of credit hours that are assigned for research course credit.

EVALUATE YOUR CURRENT COMMITMENT LEVEL

College is stressful, and maintaining the delicate balance among your academics, extracurricular activities, social life, and possibly also working at a job is challenging. And despite your best efforts, you might have a semester or two where you end up being overcommitted in several (or all) areas of your life. If that happens often, waiting to register for GPA-based research credit might be your best move. One undergrad contacted us in a panic because they had registered for GPA credit for research, fell behind in their classes, and subsequently spent most of the semester studying during their research hours. Unfortunately, their in-lab mentor, who was unaware that the student had signed a research contract, agreed to this arrangement without first checking with the PI. Then when it was time for the student to write their end-of-semester research report, a requirement for their grade, they were unable to do so because they had not done enough research.

After you know all the requirements involved with registering for research credit, evaluate your previous semesters. Consider whether you could truthfully answer yes to any of the following questions:

- Did I drop or withdraw from classes after the initial drop-and-add deadline because I was overcommitted or ran out of time to study or complete class assignments?
- Was I stressed *often* about not having enough time to complete assignments, study, or spend time with my friends?
- Did I spend the majority of my study time cramming before an exam or rushing to finish most out-of-class assignments before the deadline?
- Did I regularly ask for extensions on assignments or papers or pretend to be ill to postpone exams?
- Do I regularly feel overwhelmed with the amount of information from my classes, or can't remember much of what I read or am not sure that I'm learning anything?

If you answered yes to any of these questions, avoid registering for course credit the first semester you participate in research, if possible.

Then implement the scheduling system we cover in chapter 4, in the section "Step 1: Schedule and Prioritize Your Time," because if your current time management system isn't working *for* you, it's working *against* you.

RECONNECT WITH THE PI, IF NECESSARY

Ultimately, whether registering for GPA credit is best for you will depend on your short- and long-term goals and how well you're currently managing your academic and life balance. But another factor to consider is your financial aid package. If you register for research credit, you'll pay tuition for those credit hours, but if you register for the kind of 0-credit option described earlier, then typically the research is considered a tuition-free "class." After you have a solid understanding of your options, you might decide that it would be better to register for GPA credit, or not, and you might need to revisit the conversation with your PI. This is not disrespectful—it's being an advocate for yourself.

To start the conversation, either email your PI or say to them directly, "I would like to register for *X* research credits in the *Y* semester because it will help me reach my academic goals. Will that be possible?" If the PI responds that it's not possible, carefully evaluate their reasons. They may share something that you hadn't considered or why their approach has been successful for most undergrads they have mentored.

SUMMER UNDERGRAD RESEARCH PROGRAMS (SURPS)

If the opportunities to conduct undergrad research on your campus are limited, or you'd like to try summer research in a different place (or want to join a specific program hosted on your campus), consider participating in a summer undergrad research program.[1]

Many SURPs are designed to enhance educational equity and diversity for undergrads who are members of communities that have been underrepresented or underserved in STEMM disciplines, including Black, Indigenous, Hispanic/Latino, United States military veterans, first-generation college students, and students from backgrounds with low socioeconomic status. Perhaps the most well-known summer programs in this category are research experiences for undergrads (REUs), which are funded by the Na-

1. If it's something you want to do, we recommend participating in at least one SURP during your undergrad career or spending at least one full-time paid summer working in a lab even if it is unaffiliated with an organized program. A full-time experience will present distinct challenges and opportunities for growth that are typically unavailable through a part-time experience conducted only during the academic term.

tional Science Foundation (NSF).[2] Funding for these programs is awarded to an academic department (or multiple, collaborating departments) in a college or university, or to a museum, zoo, botanical garden, privately held research institute, or other organization. However, although REUs might be well known, there are numerous high-quality SURPs to consider that aren't funded by the NSF, so sorting through the opportunities to decide which programs to apply to can require a bit of time. As you explore the various programs, you'll want to consider multiple factors, such as where the research will take place (at a field site or in a lab, for example), how much you'll be paid, if there is a food *and* housing allowance, and whether the research offered through a particular program is exciting to you. Students who wish to participate in a SURP outside academia should also consider options with industry, nonprofit organizations, and other nongovernment institutions.

Even though one program might be promoted as an internship and another as a fellowship or scholars' program, most SURPs start in late May, June, or July; last eight to twelve weeks; and require participants to make a full-time commitment (approximately forty hours per week) to a research project during that period. As a SURP cycle wraps up, participants typically present their research as a poster or short talk on-site. Some SURPs also require participants to attend and present their summer research project at a scientific conference months after the program ends.

Some SURPs include course credit for research or an accompanying class, whereas other programs focus solely on training and mentoring students in research. Most programs include social events, seminars by scientific experts, and professional development workshops that include a range of topics, such as the ethics of data collection, how to read scientific papers and conduct literature searches, and factors to consider when deciding between grad school and other career paths. Within this framework, some SURPs are designed for students who are on a pre-health career path, while others are created for students who intend to pursue a PhD after earning their undergrad degree. These are only some examples of the customized programs offered by SURPs.

The actual research experiences offered through SURPs are as varied as there are research groups. In one experience, a student might

2. *Well-known* in this case means that PIs, grad students, and most professional scientists are familiar with NSF and REUs. If you didn't know what NSF or REUs were before reading this, don't panic—you're in good company. Most undergrads who haven't participated in research (and even many who have) are unaware of NSF and REUs. Also, note that some scientists use the designation REU as a generic term for all SURPs, including those not funded by the NSF. In this book, we use REU to refer specifically to NSF-sponsored programs, and we use SURPs to refer to all programs.

live at a field station and conduct research in conservation biology, wildlife management, or microbial ecology. In a wet lab research experience, an undergrad might work in the areas of nanotechnology, chemical biology, or biomedical engineering. And in a dry lab experience, their project might be on the topic of computational neuroscience, bioinformatics, or structure-based drug design. Even when considering these diverse categories, remember that a research project might be a combination of wet and dry lab research or field and wet or dry bench research.

How much will it cost to participate in a SURP?

Fortunately, you can be paid to participate. Although not all SURPs pay students to complete a summer internship, all NSF-sponsored REUs do. The type of paid benefits varies by program and can affect future financial aid awards, depending on how the stipend from the program is classified.[3] **In addition to being paid to conduct research, some programs cover room and board (or only cover board but provide a meal allowance) and travel to and from the institution hosting the program.** This is relevant because if you live in Nebraska but the research you're interested in conducting is happening in Washington, you can still consider a program without having to worry about cost of a plane ticket. But you'll still need to know if travel is wholly covered or only a certain dollar amount is awarded because some stipends cover only a portion of the travel expenses. Also, when attending a scientific conference is required, the costs of travel to a scientific meeting, conference registration fees, and hotel often are covered by the program, but you should confirm that so there are no surprises later. You won't get rich by participating in a SURP, but, if the stipend is substantial enough, your summer away from home might be quite affordable and provide some money for the next school term.

Can I participate in a SURP more than once?

Maybe. It depends on the eligibility requirements of the program(s) that you apply to. Some programs won't consider candidates who have partici-

3. Check with your campus Financial Aid Office to determine what impact, if any, participating in a specific SURP will have on your financial aid package. Also, be aware that the income you earn from a SURP might be considered taxable income by the United States government, and the institution that hosts your SURP might not withhold any taxes from your paycheck.

pated in any SURP, others *prefer* students without prior SURP experience, and still others don't use prior participation as a criterion to determine eligibility. We've noticed a few programs that want the same students to participate two summers in a row, but such opportunities are less common at the time of this writing. If you'd like to participate in several SURPs during your undergrad career, then starting with your second application cycle, choose programs that don't prohibit multiyear students or ones that don't outright exclude students with prior participation. However, if you enjoyed your first summer experience, even if that makes you ineligible for a second round in the same program, consider sending an email to the PI to ask if they have funds available to pay you for an additional summer. They might or might not. If not, they might know of a similar program that accepts candidates with previous SURP experience, so it won't hurt to ask.

Will I be able to design my SURP project?

It might be possible, but there are other scenarios as well. Whether you'll have input on designing your summer project will depend on the group you join and possibly the type of research they conduct. In many biomedical labs, for example, projects for SURP students are planned before the program starts. This means that the student's research focus won't be on designing experiments but on carrying them out. This approach often provides a summer student with the greatest chance of contributing to a project and gaining valuable research skills from the start instead of the risk of potentially chasing artifacts.

If I don't get to design my project, how do I know I'll work on something that is interesting to me?

A SURP might revolve around a research topic—such as building prototypes or investigating the impact of an invasive species in a specific environment—or consist of multiple host labs that conduct research in a variety of subjects. And even though you'll technically apply to a program and not an individual lab, at the application stage, some programs allow you to select one or more research groups that you would like to work with (although they probably won't guarantee placement in those groups). Other programs, however, assign each student to a specific research group within the program based on an undergrad's application essay. Regardless of the placement method used for the SURPs you apply to, you'll use the approach in chapter 4 to determine whether the overall program is right for you to lessen the chances that you'll end up stuck in a disappointing summer research experience.

If a SURP is classified as an internship, can I receive course credit at my home institution for participating?

Maybe. Some students can receive academic credit for participating in SURP internships, but others are ineligible. On your home campus, check with your academic advisor or Registrar's Office about the possibility. If you're eligible, ensure that you're clear on what you'll need to do to secure internship credit at the end of the summer. In all likelihood, before the first day of the program, you'll need to create a plan with your summer in-lab research mentor and PI that covers the expected academic outcomes of the internships and other matters. Therefore, if you decide to pursue internship credit, contact your future PI to discuss this possibility as soon as you know you'll be working with their research group.

Why doesn't everyone participate in a SURP?

To start, many students are unaware that SURP opportunities exist or how to find a program that makes sense for them. This can be particularly true for students who haven't had the benefit of strong mentorship from someone familiar with SURPs.

Also, for some students, even when considering a paid SURP, there are financial complications. For instance, a student might use the summer months to earn money to help pay for the upcoming school year or to help support their family. This can make it essential to choose a SURP with a stipend comparable to what they would typically earn working at a job during a summer *and* provide a substantial food and housing allowance to avoid additional out-of-pocket expenses. For some students, these requirements further limit their program options.

In addition, some SURPs, such as those funded by the NSF, require applicants to be a United States citizen or a national or permanent resident to participate. A student who doesn't meet these eligibility requirements won't be considered for those NSF programs.

Next, most SURPs are highly competitive, and they accept only a fraction of candidates who apply to participate. The numbers vary, but it's not unusual for some programs to receive five hundred or more applications in a single round. Some programs boast that they receive more than one thousand applications per admissions cycle. Many programs host small cohorts of students—around eight to twelve people per summer—which is significantly fewer than the number of applications they receive.

And although there isn't a standardized application for SURPs, the majority require effort and time to write a goal or research statement and secure at least one recommendation letter. Unfortunately, this is enough to dissuade some students from applying—especially if an undergrad doesn't feel that they know a professor well enough to ask for a recommendation letter or they have a hard time finding time to apply to programs in between classes, a job, and other obligations. Along the same lines, some students start searching for a program after most application deadlines have passed or when the due date is too close to put together a competitive packet in time.

When should I apply to a SURP?

One of the mistakes that students make when applying to SURPs is waiting too long to get started on the search and application processes. It's easy to underestimate the time needed to identify programs that inspire you, connect with a professor or two about recommendation letters, write a personal or goal statement (often required), and finish completing the rest of the application. And because there is no one-size-fits-all summer internship experience, you'll want to consider the expectations, advantages, and eligibility requirements of each program individually before deciding which to apply to.

Ideally, start your search for programs no later than the September *before* the summer you wish to participate. This might seem ridiculously early, but it's not. Application deadlines are often months before the program's start date, and some due dates are as early as October for the upcoming summer. If you start your search early enough, you'll have plenty of time to thoroughly review what a specific program has to offer and submit a polished and competitive application.

How many SURPs should I apply to?

This can be a difficult decision. Typically, by the time candidates are informed if they've been accepted to a SURP, the application deadlines for most other programs have passed. Therefore, applying to multiple programs around the same time is a good strategy. (On the positive side, at the time of writing this book, we were unaware of any stipend-based programs that charged an application fee.) But we recognize that, for whatever reason, you might need to limit how much time you can spare for this task. When we've conducted polls on social media about how many summer programs undergrads typically apply to each cycle, feedback ranges from as few as one to as many as twenty. But, **without knowing your circumstances, we can't advise you on what the magic number of applications would be for you.** This is a good discussion to have with a mentor who knows you well—perhaps an academic advisor or a professor you plan to ask to write a recommendation letter in support of your applications.

Can I work at a job, take classes, or take an MCAT prep course while participating in a SURP?

Probably not. The vast majority of SURPs prohibit participation in other substantial activities for the duration of the program. Such a prohibition ensures that the students they accept are fully committed to conducting the research and extracting all the benefits out of the program. Some programs do allow limited outside activities—perhaps up to ten hours per week to work or a single, three-credit-hour nonintensive class during the summer—but other programs don't support any additional activities. A prep course for the Medical College Admissions Test (MCAT) would likely be too demanding to combine with a SURP.

My family is taking a summer vacation during the SURP. Can I go with them?

Can I postpone the start date to take a break between the end of the semester and before starting the SURP?

My college is on the quarter system, so the SURP start date is before classes end. Will the program make an exception for me?

In any of these three instances, it's highly unlikely. The vast majority of SURPs expect student researchers to arrive on campus or at the field station, as applicable, no later than the first day of the program and remain committed to their research project every weekday (and possibly some weekends) for the duration of the program. So in most cases, if you can't meet a program's start date, you won't be eligible to participate in it.

It's important to carefully read the stipulations of each program you plan to apply to. Some SURPs host mandatory social or other extracurricular events during certain evenings or weekends, and others note that students should be prepared for some flexibility in their scheduling depending on the parameters of their research project.

WHY RESEARCH POSITIONS ARE COMPETITIVE (AND WHAT YOU CAN OR CAN'T DO ABOUT IT)

Depending on your major and career path, you may have already been told numerous times how important it is to get involved in undergrad research. But if an undergrad research experience is so important (and it is), and has so many potential benefits (which it does), why it is so difficult to find a project compatible with your long-term and short-term goals? In other words, why are research positions competitive? The following reasons apply to most undergrad research positions.

FIVE REASONS RESEARCH POSITIONS ARE COMPETITIVE

1. **Training and mentoring an undergrad researcher requires effort and time from your labmates and often one person in particular.** Beyond teaching fundamental techniques, a significant amount of training goes into mentoring and working with an undergrad researcher. Teaching someone how to think like a scientist and guiding them through procedures and experiments takes time and effort. The development you receive from your research experience comes from the effort you make, but it also relies on a mentor who is committed to you and your success. Just as you hope that your investment in research will pay off with professional and interpersonal development and a strong recommendation letter, your mentor hopes their training will pay off in results or data that eventually increase their productivity. Therefore, mentors try to select students who genuinely want to be part of the mentor's research program, students who have enough time in their schedule to participate, and undergrads who demonstrate an authentic eagerness to learn. These are all things that you can demonstrate by using the strategies we've provided in this book.
2. **The number of students who want to participate in undergrad research typically exceeds the number of available positions.** In essence, there is a limited number of faculty members in each department, and that number is much smaller than the number of students searching for research positions. And not all faculty members in a department mentor undergrad researchers. One physics professor shared that they don't mentor undergrads in research because the professor's work requires graduate-level math training. However, another physics professor proudly mentioned a summer research program that incorporates projects

for undergrads *and* senior high school students. Both research programs do important work, but they take different approaches. Yet, even when a professor is committed to mentoring undergrad researchers, there is a limited area of physical lab space assigned to them, so they can't always offer a spot to an enthusiastic undergrad. Another factor might be when all the members of a research team are fully committed to their current work responsibilities so that no one is available to mentor a new student. Because there is nothing you can do about faculty hires or number of available positions, it will become frustrating if you spend too much time dwelling on it. Instead, focus your attention on other matters that you can control.

3. **Eligibility requirements vary by position and mentor's expectations.** Some research mentors require prerequisite coursework, a specific academic level, or that an undergrad's research schedule be compatible with theirs. Some SURPs accept students only from institutions that have limited research opportunities, students who are early in their academic career, or students who are considering attending grad school in the SURP's topic. You won't match all eligibility criteria for all positions, so you'll be ineligible for a certain subset of them. Again, there isn't much you can do about this, but you can save effort by applying only for positions for which you're eligible.
4. **It's often the student who knows how to navigate the hidden curriculum when writing emails who is offered an interview and position.** An email is an excellent way to secure a research interview—if you know how to navigate the hidden curriculum when you write it. An interview presents the opportunity to demonstrate enthusiasm in the interviewer's research and an elevated level of professionalism—which, again, is influenced by the academic hidden curriculum. However, certain negative impressions included in an application or mentioned at an interview can eliminate a student from consideration even if that undergrad would have been an ideal addition to the research team. We cover common ways students send negative impressions to potential mentors both in email and in person in chapter 5.
5. **It's a matter of timing. An eligible student who applies for a research position at the "right" time can lock down a position.** The problem is that you won't know what the "right" time is for a particular mentor. One mentor might select students at the start of the semester, another near the end. Some mentors don't review research inquiries until four to six weeks into the semester, and others consider applications midsemester and then invite undergrads to join their research group as an observer (but not officially start research) until the start of the next semester. It's safe to presume that academic breaks aren't a popular screening time for

most. The solution is to use the First Contact tips we provide in chapter 5 to emphasize genuine interest and enthusiasm. This doesn't guarantee that a potential mentor will have an available opportunity when you make contact, but it does increase your chances of being offered an interview if they do. And if you're applying to a SURP, start your search early in the fall semester so you don't accidently miss the deadline.

Unfortunately, you can't change the fact that research positions are competitive. However, there are strategies to present yourself as a competitive applicant. The rest of this book covers those strategies, including the most common mistakes undergrads make, so that you can avoid making them yourself.

PART TWO

4

Your Search Strategy

A successful search for a research position is about more than reading project descriptions and deciding whether to apply. It's also about determining how much time you realistically have for a research experience and keeping your search organized so you don't become overwhelmed during the process. Additionally important is understanding the most common mistakes others make so you can avoid making the same errors.

TEN SEARCH MISTAKES TO AVOID

1. **Not making your search a priority.** It's tempting to put off the search for a research position, but doing so won't make it any easier to get it done. As soon as you know you want to participate in undergrad research, make your search a priority. **Be aware that it can take substantial time to identify an experience compatible with your goals in a research group that has an available position.** Start your search early in your college career—ideally early in the semester *before* you want to begin conducting research. You might be fortunate to find a project quickly, or you might need to spend half a semester or more searching. If it's already late in the semester and you want to start your research experience shortly, don't worry, but start your search as soon as possible.
2. **Using an ineffective (or no) scheduling system for your academic, job hours, and personal and social lives.** Prioritizing the activities that matter the most to you exponentially increases your chances for succeeding in those activities. And using an effective scheduling system will help you avoid overcommitting yourself to some activities at the cost of others. If

you **determine how much time you have (and are willing) to dedicate to a research experience at the *start* of your search**, it will be easier to pursue or eliminate research opportunities based on your actual availability as opposed to wishful thinking. This approach also increases the probability that you'll be able to uphold the time commitment you agree to at a research interview, which is key to building mentoring relationships with your labmates.

3. **Placing too much emphasis on reading a scientific paper.** Some undergrads are advised to read scientific papers as the first step in their search for a research position. Unfortunately, the pressure to do so can derail a search before much progress is made. Reading some of the research articles by the principal investigator (PI) works for some undergrads, but it's overwhelming for others. For some disciplines, a student with no academic background in a topic will have little hope of understanding the title of a paper, much less the methods, results, or conclusion. **It's advantageous to try to read a paper, but if it's too frustrating, don't let it become a reason to delay or give up on your search for a research position.**
4. **Not learning something specific about a research program.** Your ultimate goal should be to find a research experience that will become a meaningful and rewarding use of your time. Many students, after struggling to read a scientific paper, become overwhelmed and give up trying to learn about a PI's research program. But **to ensure you have a genuine interest in the research, and to submit a more competitive application, you'll need to learn *something* about a specific project or potential mentor's research focus before applying to conduct research with them.** Fortunately, there are multiple ways to learn about a PI's research that don't involve reading a scientific paper, such as reading on the internet about the PI's research interests, reading from a scientific poster hanging in the hall outside the lab, or reading an advertisement for a research position. We cover all these approaches later in this chapter.
5. **Only considering a subset of inspirational words as relevant.** The description of a research project or program could inspire you to consider an opportunity, but it shouldn't be used as the *sole* factor to *determine* whether the experience will be a meaningful use of your time. Your research experience will be much more than the title of a research project. The specific opportunities offered by working with a certain research group or in-lab mentor are equally relevant, especially if any of your goals include professional or interpersonal development.
6. **Trying to replicate a classmate's research experience.** Even if you accept a position working with the same research group as a friend does, you won't have an identical research experience. Find inspiration from your classmates, but remember that your research experience will be guided by your goals, the effort you invest in the opportunity, your specific project,

and the relationships you develop with other group members. We present a strategy with specific questions to ask your friends about their research experiences later in this chapter.

7. **Relying only on advertisements to find a research experience.** Advertisements for undergrad research positions posted on department websites or social media pages are useful when they work. But advertisements quickly can become outdated or filled after only a few days. In addition, not all mentors with available projects advertise for undergrad researchers in these venues. Your search might include reviewing advertisements, but that should not be your sole approach.
8. **Allowing the possibilities to overwhelm.** Considering every research group, every posted position, every discipline, every research project, and every PI's research interests will lead to frustration. **If you always think the grass is greener somewhere else, then you'll never get started on a research project, or you won't stick around in a research experience long enough to accomplish much.** Instead, it's best to consider each research opportunity individually and periodically evaluate whether the grass is green enough after you've begun your research experience.
9. **Becoming emotionally attached to an unobtainable position.** When reading the description of an available research project, an incredible summer research program, or a research group's collective approach to mentoring undergrads, you might deeply connect with the words and feel that no other experience could possibly match it. But if that position includes *eligibility requirements* that you don't have, or there isn't currently an available position, then let it go. As an undergrad, you'll discover that there probably isn't "The One" project or research experience that is perfect for you—there are several. Letting disappointment close you off to other possibilities only serves to further limit your opportunities.
10. **Not implementing a search strategy.** A search strategy is a much better approach than randomly picking a research group and hoping the experience works out or accepting any offer just to be done with the search. A search strategy doesn't have to be complicated to be effective—for most cases, it's a matter of breaking up the task into small, achievable steps. This chapter provides a detailed strategy to complete that process. **Using a strategy-driven search will save time and will help you find opportunities compatible with both your personal and academic goals.**

GETTING STARTED IS THE HARDEST PART

For most students, the mere thought of searching for a research position is enough of a reason to put it off. There are several reasons for this "failure to launch." First, the search feels like a huge chore. Second, it can be overwhelming, as the term *research experience* encompasses a seemingly endless number of research groups and projects, and it's a challenge to decide which to pursue. Third, without an organized search strategy, the process can quickly become a lot of work for little gain. Add in the standard advice to read a scientific paper before contacting a PI, and it can be difficult to remain enthusiastic and find the time to prioritize a search. However, it would be a shame if any of these reasons prevented you from getting started because a research experience could be a defining aspect of your time as an undergrad.

The purpose of the search strategy detailed in the sections that follow is to accomplish two key steps in the least amount of time possible. These steps are **(1) determine how much time you can dedicate to research;** and **(2) identify potential research opportunities that inspire you.** This strategy is designed not only to save you time but also to be a building block of your research experience. For instance, carefully examining your schedule will help you find a research experience you realistically have time to participate in as well as maintain your academic and life balance after you join a research group. In addition to learning how to prioritize your time, you will find that these approaches will help you uphold the commitments you make to your research mentor, which will in turn make direct, positive impacts on the professional mentoring relationship you develop with them.

If you follow the search strategy described in this chapter, the hardest part of your search will be finding the self-discipline to get it done. *You can do that.*

STEP 1: SCHEDULE AND PRIORITIZE YOUR TIME

Finding the right academic and life balance is tough, and it's harder to do than anyone imagines it will be. But the ability to plan your academic schedule so that it coexists with your college experience and life is essential to both your happiness and your success. To do this well, you'll need a solid time management plan, self-discipline, the ability to prioritize, and the courage to make some tough decisions, such as letting go of low-value activities. If your plan is to participate in a full-time summer undergrad research project (SURP), you won't need to consider your schedule with the same level of detail as described next, but you should still schedule time to identify programs and complete applications. Additional tips on maintaining your academic and life balance and connecting with campus wellness resources are presented at the end of this chapter.

PART A: CREATE YOUR SCHEDULE IN A CALENDAR APP

Your goal for this part is to answer the question, "What do I do all day?" If you're not already using a digital scheduling system or you're using a paper-and-pen-based approach, we recommend switching to one of the many free, web-based calendar apps as they are feature-rich and easy to use. (Google Calendar is what we use, but there are many others to choose from.) Ultimately, you should be able to schedule all your time commitments (classes, exams, job hours, volunteer activities, planned social events, etc.) and display them in a variety of formats. Using an online app, you can access your calendar across most platforms (smartphone, tablet, computer), schedule new activities, make changes to existing ones, and set reminders to help you remember the important activities or deadlines. A digital calendar also makes it easy to reschedule activities and ensure that none conflict, which is an invaluable feature when adjusting your schedule.

To start, create your current schedule in your preferred app. For this to work, you need to be completely honest with yourself about how you currently spend your time—not with how you *should* spend your time. To stay focused, consider installing an app on your phone and computer that temporarily blocks your access to distracting websites. (At the time of this writing, Cold Turkey and Focus are our top recommendations, but they might not be available when you read this, or you might prefer to use another app.) Then settle in with a latte and start your favorite music playlist. Pick a typical week in the semester to get started, or fill in the blanks with what you did last week. This step doesn't need to be perfect, so don't overthink it or get stressed about where your time goes while you do this. Estimates are fine. You'll make adjustments as needed in part B. If you're a parent or a caregiver for children or someone else, you might wish to start by blocking out time that you're unavailable because of those caregiving obligations unless the research position has a virtual element.

Assign a different color or label to each of the following time commitments as you add them to your calendar:

- *Class schedule*
- *Work or internship schedule*
- *Study time* (include time spent working on assignments)
- *Extracurricular activities* (clubs, volunteer work, doctor shadowing)
- *Family commitments*
- *Sleep time* (Yes, you need to add this. You're tracking *where all your time goes*, and sleep is part of it.)

And everything left after that is . . .

- *Personal time* (social activities, show streaming, exercise, video games, and mundane life chores such as laundry and, if you're not living in a residence hall with a dining plan, grocery shopping and meal preparation and cleanup)
- *Commute time* (add if you spend a substantial amount of time commuting to and from campus.

PART B: EVALUATE YOUR SCHEDULE AND MAKE ROOM FOR RESEARCH

The standard approach used to determine how much time to devote to a research experience is to ask how much time is *required* and then try to make it work. Unfortunately, this strategy relies heavily on wishful thinking instead of solid planning. For some students, it leads to overcommitment; yet for other undergrads, this approach results in an immediate rejection of a perfect research opportunity because the idea of *X* hours per week feels overwhelming. A better approach is to **determine the amount of time you can dedicate to a research experience and then explore opportunities compatible with that time allowance.** This will enable you to consider or rule out a position based on your available hours, which lessens the chance that you'll become overcommitted in a position later or mistakenly rule out an option because it *seems* like too many hours but would not have been. Knowing how much time you can reserve for research is always preferable to estimating a commitment level without giving it much thought.

To determine your ideal research schedule, first return to chapter 1. Using the sections that you highlighted as a guide, determine the approximate weekly time commitment you'll need to accomplish your professional research goals. Call this estimated commitment "research hours" and block out the time in your online calendar. But remember, this is only a placeholder to help you get started and issue a reality check on how much time you have for research and determine whether that time commitment aligns with your current goals. In the future, you'll consult with your in-lab mentor about your official schedule.

As you create your calendar, also keep in mind how you'll commute to and from the research site. If you rely on taking a bus, for example, or don't have safe transportation after dark, take those factors into account as you create your potential research schedule. You might need to consider projects that can be done virtually, or include a virtual component, when coordinating your schedule.

If it's not immediately obvious how to incorporate research hours into your calendar, rearrange the flexible areas of your schedule to make room. Determine where you can make changes, where you can move or reduce hours, and which activities you can remove altogether. Evaluating your priorities and adjusting your schedule to reflect them constitute an important step in personal development because knowing what activities you don't have time for is equally important as knowing what you do have time for. **Purposefully reflecting on your activities and obligations is how you learn to organize one of your most important, and limited, resources—your time.**

To make room for research, consider the following questions about your current time commitments and make adjustments where you can. We don't include adjusting a work schedule among the items that follow because those hours are often beyond someone's control. If, however, that is an option for you, then consider adjusting your work hours as well.

Personal Time

Can I reduce the amount of time I stream shows? Can I prioritize hanging out with my friends on the weekends or at night to make space for research during the day? Am I getting the most out of the leisure time I have, or could I give some of it up to pursue a meaningful research experience?

Extracurricular Activities

Can I schedule volunteer activities on a different day or time or reduce the number of hours I participate each week? Am I getting enough out of each activity to justify continued involvement, or would I be willing to give up something to free up time for research? Am I involved in activities that are important to me, or do I participate in them because it's important to someone else? Can I combine any extracurricular activities and personal time?

Class Schedule

Can I drop or add a different section at an earlier or later time? Would taking some classes online work for me?

Sleep Time

What is the number of hours I need per night to excel? If I get more sleep at night, could I reduce the number of naps I take during the week? If I increase my sleep time even a small amount, will my study time be more effective?

Study Time

Can I study more hours on the weekend? Could I study during my laundry cycles instead of scrolling through fun stuff on my phone or streaming shows? Do I have time in between classes that I could use to study or work on class assignments? Are the classmates I study with committed to the study sessions, or does the "study" time almost always turn into personal time? If I commit to a study schedule without allowing interruptions, will I be more productive overall? Could I benefit from installing internet site blocking software on my computer to use when I study to make that time more productive?

As you reflect on the proposed research hours now on your schedule, how do you feel? Imagine that your research experience will be a meaningful and rewarding use of your time. Will you be able to meet the time commitment? If yes, then you have a good starting number. **If you feel that the sacrifices you'll need to make are too great, go back through your schedule until you determine a number that will work.** This is also the time to evaluate, especially if you're working at a job *and* managing college responsibilities, whether you'll only be able to join research experiences that are paid or include course credit for participating.

KEEP YOUR COURSEWORK ON TRACK: A NOTE ON WHY IT'S IMPORTANT TO SCHEDULE STUDY SESSIONS

You shouldn't sacrifice your academics to participate in a research experience; however, this doesn't automatically give you license to neglect your research responsibilities each exam cycle. If your current approach to studying is to cram a day or two before each exam, you need a new system. The reason, beyond keeping your academic and life balance intact, is that the more last-minute "stuff" you must do, the more likely it is that you'll cut your research hours to get things done. On the surface, this might seem like a good solution, and on occasion it might not create an issue. However, this approach could lead to unintended consequences, such as limited opportunities to learn new things in the lab. Therefore, ***schedule* your study time and use that time well.** If you do need extra hours to study or complete an assignment, *try* to take them from your personal time—not your class, sleep, job, or research schedule. For specifics about why continually cutting your research schedule can be risky, review the chapter 3 section "The Time Commitment."

PART C: DETERMINE THE MAXIMUM NUMBER OF HOURS PER WEEK THAT YOU CAN GIVE TO RESEARCH

The last step is to determine the maximum hours per week you're able (and willing) to commit to a research experience with your current schedule. **Knowing your absolute maximum commitment level will prevent**

overcommitment because you'll be less likely to pursue a research position that requires more hours than you realistically have available. It will also be important to know this number at an interview when the required number of hours per week is discussed. And if you're applying to SURPs, it's important to realistically evaluate any additional activities you plan to do during the summer because, as you're aware, most programs will expect a forty-hour work week commitment.

STEP 2: IDENTIFY POTENTIAL MEANINGFUL RESEARCH EXPERIENCES

You might be wondering, "Why does it matter if I care about the research project I work on? Wouldn't it be easier to take the first opportunity I'm offered?" But how you spend your limited time in college does matter. Research is challenging. It often includes boring, frustrating, or repetitive aspects, and participating costs time that could be spent doing other activities. For it to be worth what you give up to participate in undergrad research, you need to care about the project on some level or like doing most of the procedures that make up the core of the experience. And **if you join a research project where you're genuinely interested in the subject matter, it will make the challenging moments easier to get through and your overall experience more rewarding.** Also, you'll be more likely to engage with your project on a creative level and generate new ideas or recognize when something is odd that might warrant further investigation. In some research experiences, taking these actions will be expected by a student's in-lab mentor and PI and be required for the life of the professional mentoring relationship.

After you have a realistic idea of how much time you can dedicate to a research experience, you're ready to search for opportunities that pique your interest. To help you identify opportunities in the rest of this chapter, we cover some traditional search approaches and other options. If you're mostly considering SURPs, you'll identify most opportunities by using the classic approach of web and database searches.

PART A: DOWNLOAD A REFERENCE MANAGER TO ORGANIZE YOUR SEARCH RESULTS AND OTHER MATERIALS

Most undergrads we've asked used the internet at some point in their search even if they visited a career center or undergrad research office. However, to be the most efficient, you'll want to do a little preparation before starting an internet search for potential research experiences.

As you identify potential research opportunities, you'll want to collect and organize them in a single, digital place. You'll want to add new information easily whether it's text clipped from the internet, or links to important websites, or journal articles, or photos of research posters. Then, when you're ready to start the application process, you'll want to access all these resources from any of your electronic devices. We recommend using a reference manager for this task.

Most scientists use a *reference manager*, also sometimes called *reference software*, to keep track of journal articles, books, and other reading materials related to their projects and field. These programs have a variety of features, but many include free versions that allow annotating and highlighting within a document, which are key tools for an undergrad searching for a research position. If you're already familiar with a reference manager (perhaps it was required in a class), continue using that one for your search. If you've never used one, do a web search for "free reference managers" and pick one. Alternatively, choose Zotero or Mendeley, both of which had free versions available at the time of this writing. Plus, both those programs make it easy to clip materials from the internet after installing a plug-in on your browser.

While using a reference manager, either create folders or use tags with keywords such as "ads for research positions," or "Dr. X's research interests" to make information easy to find later. Even if you're primarily searching for SURPs, using a reference manger to track which programs you're excited about, and the research groups you hope to join within those programs, will help your search stay organized. That will eventually translate to spending less time on the search process.

At the start, learning to use the reference manager software might be frustrating but only because it's new. If you get stuck, check for tutorials on the software's website or do an internet search for "how do I do *X* with [name of reference manager]." It's worth your time to learn how to use a reference manager. When you join a research position, you'll use it to keep track of journals articles you plan to read, record your thoughts as you read them, create bibliographies, and possibly collaborate with labmates in a variety of ways. **Any experience you gain from using a reference manager before you start a research project becomes an advantage after you start a conducting research.**

PART B: CUSTOMIZE AN INTERNET SEARCH AROUND YOUR INTERESTS

A web search is quick and will provide seemingly endless possibilities to consider. However, that's also how an internet search can derail your self-discipline. The easy part is getting search results—the hard part can be sorting through the massive number that are returned in 0.5 seconds to decide which ones are right for you.

When sorting through the results of a web search, you'll need to strike a balance between using inspirational words and considering every possible research group and opportunity that pops up. To do this, avoid trying to distinguish between complex research descriptions, such as those for research programs that study the neurogenetics of behavior in *C. elegans* or the epigenetic modifications in cancer cells. The fine details will only distract you and make your search more difficult. Instead, ask yourself if one or both of the research programs are interesting enough to pursue a research position working with the associated group.

One way to determine whether you connect with a research program or topic is to focus on how you feel when first reading about it. **At least one of the following three things should be true before you consider a research group for your experience:**

- You like the discipline. Examples: microbial ecology, genomics, computational biology, physiology, biochemistry.
- The research program uses the same or similar techniques that you enjoyed using in a lab course or liked learning about in a lecture course.
- You share the PI's research interests, or a defined available project seems interesting.

To identify research opportunities or specific research groups you might like to work with, we recommend using up to five customized phrases for your web search and then spending a modest amount of time sorting through the results. You probably won't need to use all five—but if one doesn't give you the results you desire, move to the next. Keeping your search simple is the key to keeping it efficient. This approach should take you less than an hour to identify ten to fifteen potential opportunities that are interesting to you.

Customize the first set of web search phrases to reflect the name of the college or university you attend. Basically, you might think, *This [topic, field, technique] is something I'm interested in, I wonder who on campus studies it?* and then use that prompt to customize web search phrases. For example, if you attend the University of Minnesota, major in genetics, and are interested in drug discovery research, you might do a web search with the terms "University of Minnesota undergrad research opportunities" or "University of Minnesota genetics undergrad research" or "University of Minnesota drug discovery research." In many cases, you'll find more than enough opportunities to explore with one to three of the search terms. Also, you can add "paid" to any of the search phrases listed next, but know that it might not return all (or any) funded positions.

Web Search Phrases for Research Experiences on Your Campus

- [Your college or university] undergraduate research opportunities
- [Your college or university] paid undergraduate research opportunities
- [Your college or university, your home department] undergraduate research opportunities
- [Your college or university, your major] undergraduate research opportunities
- [Your college or university] [research topic][1]
- [Your college or university] undergraduate research scholarships (or fellowships or internships)

Web Search Phrases for SURPs

Use these phrases to identify SURPs—including research experiences for undergrads (REUs) offered through the National Science Foundation (NSF), as described in chapter 3—that align with your interests. You might have less luck if your institution has limited research opportunities and you include its name in the search. However, it only takes a few minutes if you want to try the search both ways.

- Paid summer undergraduate research programs
- Paid URM summer research programs (URM meaning *underrepresented minorities*)
- REU programs
- Undergraduate research national lab
- Paid summer undergraduate research programs at [your college or university]
- Summer undergraduate research fellowship in [topic or field]

Search for SURPs Hosted by Specific Scientific Organizations

Another search strategy is to identify SURPs that are hosted by a specific scientific society that is reflective of your desired career path. For example, the Society for Developmental Biology (SDB) has a competitive summer scholars program for undergrads who are interested in eventually

1. Research topics could include a specific disease (Alzheimer's disease), a research organism (*Drosophila*), a discipline (evolutionary biology), or your own personal research-related interests.

attending graduate school in developmental biology or a related topic. So if you're genuinely interested in developmental biology, you would want to explore the SDB website (https://www.sdbonline.org) for potential programs. This type of search can be tricky because if you don't know that a scientific society exists, it's difficult to know to search for programs it might sponsor. Therefore, ask any professor or graduate TA for suggestions on scientific societies in the subjects you're interested in and search the internet with [scientific society in *X*], where *X* is the subject you're interested in.

Search for Databases of SURPs

Another option for SURP candidates is to search a database of curated opportunities (including listings for NSF-sponsored REUs). Although databases that host on-campus opportunities can be problematic (as we will discuss in a later section), ones for SURPs are frequently updated with wide-ranging opportunities. No SURP database is all-inclusive, but each repository can be a valuable resource during your search. For example, PathwaysToScience.org (https://pathwaystoscience.org) has a search function that allows visitors to browse NSF programs by academic level, start date, discipline, and other factors. The NSF website (https://www.nsf.gov) also hosts its own database of REU sites where it's easy to sort opportunities by keywords or browse by subject. The Ecological Society of America (https://www.esa.org) posts opportunities for summer jobs, internships, and REUs. The American Society for Biochemistry and Molecular Biology (https://www.asbmb.org) maintains a database of some summer internships for undergrad researchers, and you don't need to be a member of the society to use it. The Society for Advancement of Chicanos/Hispanics & Native Americans in Science (https://www.sacnas.org) lists REU programs as well as numerous other resources for students and professionals in science, technology, engineering, math, and medicine (STEMM). And if you're interested in the vast topic of plant biology, the American Society of Plant Biologists website Plantae (https://plantae.org) lists internships by categories such as ecology and evolution, computational biology, and general biology, which you can then use to identify individual programs of interest.

This is only a partial list of places to find SURP databases.

PART C: EVALUATING THE WEB SEARCH RESULTS AND MOVING PAST FEELING INTIMIDATED

The suggested web search terms or a visit to one of the databases previously mentioned might produce the following:

- Advertisements for undergrad research positions on your campus (these might be associated with a campus-wide database or a listing curated by a specific department on campus)
- Advertisements for SURPs, including NSF-sponsored REUs
- Links to campus undergrad research centers or other resources such as college-wide research opportunities
- Departments, research institutes, colleges, or centers related to the topic of your search
- PIs' web pages or links to their faculty research interests

Regardless of what you read—an advertisement for a research position, a PI's research interests, a scientific paper, or a lab's SURP description—don't become discouraged or intimidated by the scientific terminology. At this point, you're only using these resources as a starting point to explore your interests and consider the possibility of applying to one or more of these positions or research groups. Here, your only goal is to decide if the overall program, discipline, or techniques are interesting to you. You'll create a short list of positions later from these results.

The search results you receive will depend on the terms you used but similar strategies can be effective with either advertisements for undergrad research opportunities or faculty research interests, which are covered later in this chapter.

Advertisements and Databases for On-Campus Opportunities

An internet search for on-campus opportunities will return some advertisements from individual researchers (a grad student, postdoc, staff scientist, or PI) or databases hosted by a department or possibly the Office of Undergraduate Research, if your campus has one. However, these advertisements from individual mentors are not the same as the *job* advertisements described in the subsequent section "More Methods to Identify Undergrad Research Opportunities." The advertisements discussed in this section don't pertain to student employees.

Scanning campus research databases can be effective for someone who connects with a potential mentor before many of their classmates notice the advertisement—especially if the student knows how to navigate the hidden curriculum and uses the expected and enthusiastic approach at First Contact (the first time the student contacts a PI to ask for a research position). However, potential downsides include that it's not unusual for these advertisements to remain posted for months (or years!) after a position has been filled. One undergrad shared that they sent several emails in response to an advertisement until the department chair informed them that the researcher listed on the project had left the institution two years prior.

Another potential issue with on-campus databases arises when a student feels overly discouraged by an absence of opportunities that reflect their interests or feels pressured to apply for a position that doesn't align with their interests or goals because they believe that opportunities outside the database don't exist. This is unfortunate because **many research groups don't use databases to advertise for research assistants** and instead offer positions to undergrads who understand the hidden curriculum of connecting with a potential mentor in an instructional lab or after a lecture class, after attending a seminar related to a professor's research program, or other methods. Students who are unaware that these avenues exist wouldn't know to pursue them. We address using these and other methods later in this chapter. But covered next is how to maximize using advertisements and databases during your search.

Using a Database for Finding and Evaluating a SURP or On-Campus Research Opportunity

When they work, advertisements for undergrad research opportunities can be a gold mine of information. Typically, they cover basic requirements (time commitment, prerequisite coursework, academic level), the research topic, the background or significance of the research, and a synopsis of the PI's or potential in-lab mentor's research focus. Many, but not all, share whether the research can be done on campus or at a field site, or whether there is a virtual option for participants. Advertisements sometimes include whether participation is likely (or not) to lead to a publication. Advertisements for general lab assistant positions (paid or unpaid) that could morph into research positions typically list the expected time commitment in number of hours per week, specific responsibilities of the position, and work-study award requirement, if any. Mentors generally don't advertise observational positions.

At a minimum, advertisements for SURPs should list the program's eligibility requirements, start and end date of the program, where it's located, stipend amount, housing arrangements, food allowance, if any, and expectations, such as how many work hours are required per week. These advertisements also should include the overall goals of the program and a description of each participating research group's focus, written in accessible language so that it's easily understood by someone with little background on the subject.

When you read an advertisement for a research position, ask yourself:

- Does the opportunity seem like something I want to do?
- Can I fulfill the time commitment? (If given)
- Do I have the required prerequisites or meet the eligibility requirements? (If any)

If your answer to all three of these questions is yes, then save the advertisement in your reference manager.

Afterward, go through your notes and highlight the name of the project and a sentence or two that covers the main reason(s) you're interested in the position. (If your note-taking app doesn't allow highlighting, write bullet points at the top the document). Even if you're interested in everything the position description describes, highlight only the three sentences (maximum) that you connect with the most. You'll want to find this information quickly in the next step of your search, and if you highlight a whole page of text, it defeats the purpose. If your search returned a compilation of advertisements, continue through the lists until you've identified, uploaded, and highlighted sentences in ten to fifteen possibilities—or until you've reached the end of the list, whichever comes first.

Faculty Research Interests

As previously mentioned, **not every undergrad mentor advertises when there is an available position with their research group.** For instance, a PI might believe that a highly motivated student who is genuinely interested in conducting research won't rely on advertisements to find a research position. Another PI might not advertise because they receive numerous email inquiries from students who want to do research each semester, or they recruit students from their classes, or because someone else in their group handles the interview process for undergrad researchers. Also, some PIs don't advertise because they don't mentor undergrad researchers. Unfortunately, there is no way to know which, if any, of these reasons apply to a particular PI unless you have inside information about how they manage their research group.

But the good news is that none of these reasons prevents you from connecting with a PI with whom you'd like to do research. By reading a PI's faculty research interests, you can determine whether their research program seems like one you'd like to be part of and then contact the PI if it does. Most PIs maintain a web page wherein they describe their research interests, which is sometimes referred to as a *research overview* or a *faculty research description*. This is a summary of their research program, problems or questions their research addresses, and the significance of their research. Some research interests include an overview of techniques used in a research program (such as mass spectrometry, bioinformatics, or cell culture) or broad disciplines related to the research (such as organismal physiology, landscape ecology, or microbial geochemistry). Many PIs manage multidisciplinary research programs, so it's not unusual for them to list several research topics in their faculty interests.

How to Find and Use the PI's Research Interests in Your Search

If your web search turns up a department, institute, center, college, museum, or similar entity, navigate to the associated faculty research interests to find research that inspires you. If your search doesn't guide you to them, you can identify inspiring faculty research interests and determine whether the work the research group does will help you accomplish your goals by following this strategy:

1. Use the internet to visit the home page of the department, institute, etc.
2. Navigate to the faculty directory page.
3. Browse the page. When a PI's research interests, program description, or associated keywords are interesting or inspiring, save the page in your reference manager.
4. Highlight a sentence or two that covers the main reason(s) you're interested in the research. Even if you're interested in everything you read, highlight only the three sentences (maximum) that are the most important to you.
5. Repeat these steps until you have a total of ten possible research opportunities.
6. Next, visit the personal web page of each faculty member to determine which ones list undergrads as research group members, mention projects for undergrads, or highlight a current or former undergrad's research accomplishment. Consider prioritizing applying to these research programs above those that don't *specifically* mention undergrads on their website. But keep in mind that not all PIs or research groups that mentor undergrads will mention this on their web page or update it regularly, so don't eliminate a lab that inspires you based solely on this metric.

MORE METHODS TO IDENTIFY UNDERGRAD RESEARCH OPPORTUNITIES

There are several reasons you might want to use other approaches to search for a research position instead of (or in addition to) the search methods described previously. Many of these, however, rely on timing to be successful. For example, advertisements on a department's website for on-campus research positions might be outdated and an opportunity could remain posted long after the position has been filled or the researcher supervising the study no longer works with that research group. Or you might want to attend an on-campus research symposium or research fair to connect with researchers conducting studies you'd like to join, but if there isn't an upcoming one, it won't be an immediate option. And certain programs, such as bridge to baccalaureate programs, are only offered at some colleges.

In addition, not all PIs describe their current research on their websites or update their research interests regularly. When you explore research opportunities off-line, you might find an inspirational project described on a poster or learn about one at a seminar. Or you might interact in person with a researcher who shares their passion about their project, and possibly turn that interaction into an interview.

Some of the approaches described next make it easy to demonstrate that you're self-motivated, creative, genuinely interested, or willing to put extra effort into finding a research position. If it's genuine, there are no drawbacks to this.

CONNECT WITH THE OFFICE OF UNDERGRAD RESEARCH, ACADEMIC ADVISORS, AND OTHERS ON CAMPUS

On your campus, there might be a Career Resource Center, Office of Undergraduate Research, or an individual in the department of your major who can provide guidance for finding an undergrad research experience or helping you prepare for one. (We even know of a few who recommend our website UndergradInTheLab.com to their students to help prepare for and excel in an undergrad research experience.) Some campuses also have student-managed groups with peer advisors who function similarly.

When connecting with the staff at a career center or undergrad research office, you'll interact with professionals who are familiar with all types of scholarly activity on your campus, so they can be helpful no matter what your major is. Sometimes they know of specific professors who currently have open research positions, professors who have a long-term track record of mentoring undergrads in research, or they can suggest SURPs or off-campus internship experiences (paid or unpaid) for which you're eligible. And if you're searching for an experience that combines science, technology, engineering, arts, and mathematics (STEAM), they may know other students who are also hoping to make cross-disciplinary connections. Also note that the office might maintain a digital database of advertisements from researchers on (or off) your campus.

An advisor for your major or the preprofessional track you're on might have additional suggestions on where to search for research opportunities. They probably won't have a list of research groups with available positions, but again they might know which professors regularly mentor undergrad researchers. Advisors also might know of a specific database that advertises undergrad research opportunities (in your department or outside it), know about a departmental or college-wide research symposium you could attend (details on how to make this work for you are covered later in this chapter), or be able to direct you to specific research internships (on or off campus).

We do want to mention, however, that the feedback from students on the helpfulness of these resources varies. Several students, for example, found the workshops offered by their undergrad research office immensely helpful. These workshops might include how to write a research proposal or apply for funding to conduct it, effectively communicate with a research mentor, design research posters, craft a curriculum vitae (CV), and more. One student, however, was frustrated when the person they met with just did an internet search (akin to the one we recommend in the previous section) and emailed them the results. This wasn't unhelpful per se, but it wasn't the customized help that the student expected.

However, even if the person you meet with suggests a specific research group or project, it's essential that you still do your homework to determine whether it's the right opportunity for you before pursuing it. We've connected with a few students who weren't interested in a research experience suggested by a professor, teaching assistant, or someone else but the student felt obligated to contact the recommended research group anyway because the person advising them was kind and enthusiastic. (Sometimes a similar situation happens at interviews, too, but we'll cover how to handle that in chapter 6.) Sorting out whether a research position is right for you isn't as difficult as it might initially seem if you use the strategy we covered earlier in this chapter.

APPLY FOR A BRIDGE TO BACCALAUREATE PROGRAM

If you attend a community college or state college, your school might participate in a bridge to baccalaureate program with a four-year university. In particular, programs funded by the National Institutes of Health (NIH) assist members of underrepresented and disadvantaged groups who are making the transition to a four-year university to complete their bachelor's degree. These programs, often referred to as *scholar bridge programs*, provide a variety of benefits to student participants. For instance, some programs include career development workshops, social and cultural support, tuition and fee waivers, and ongoing, formal mentorship. But a key component of all these NIH-funded bridge programs is placing students in undergrad research experiences with biomedical and behavioral research groups.

The eligibility of each program varies. For some programs, a student isn't required to have been accepted to a four-year university as long as doing so is part of their career plan. Grade point average (GPA) is a consideration; many programs require a minimum 2.8 GPA and continuous participation in the program once admitted . United States citizenship or being a permanent resident or noncitizen national is required if the funding originates from NIH.

To find out whether your state or community college participates in a bridge program, do an internet search with "bridge program [your college]." If the search is successful, read about the eligibility and benefits of the program to help decide whether it's right for you. If you're not able to find a program after the internet search, connect with your academic advisor or any professor you already consider to be a mentor to help you determine whether a bridge program is offered through your institution.

PERUSE JOB ANNOUNCEMENTS

When a PI wants to hire someone for the lab team, they advertise the position through several venues, but typically you can find notices on the official job listing site hosted by the institution. These advertisements are of paid positions—*for employees*—and many students don't know to search them because they are unaware that an undergrad can also be an employee. Usually, these opportunities are for general lab assistants and pay an hourly rate with no health care or other benefits, and sometimes eligibility is linked to having a work-study award. Applying for these positions is sometimes done through an online portal or by emailing the PI directly.

CHECK BULLETIN BOARDS

Even in this digital world, announcements for undergrad research opportunities can be found on literal bulletin boards in campus buildings. As soon as you know what department(s) or discipline(s) you're inspired by, when you're between classes, check out bulletin boards for notices related to research opportunities, volunteer and paid lab positions, seminar notices, and research symposia. In addition, starting in the fall and continuing through early spring, SURPs often mail large, shiny announcements extolling the benefits of their summer program to other institutions. If you notice an interesting announcement, take a photo of the flyer and upload it directly to your reference manager app, if possible, or a folder in whichever cloud storage app you're using to store such information. Just be sure to add the photo to a folder with a descriptive label such as "Potential Research Opportunities" so that when you have the time later to consider the opportunity, it's easy to find.

PAY ATTENTION TO RESEARCH POSTERS

Research posters are an incredibly valuable resource yet are often an undervalued tool in the search for a research experience. **Researchers**

create posters for professional meetings and often display them in the hallways outside their labs or offices. Posters don't tell the entire story of a research project—they are a synopsis of a research project or one aspect of it, which is one of the reasons they are so useful. When compared with a published paper, posters are easier to read and understand, but they still have enough information to inspire and to provide the research objectives of a project. Although formats vary, a poster may or may not include the background information or significance of the research project, an abstract, techniques used, or plans for the next steps of the project. What all posters do contain, however, is a list of contributors and their affiliations. (Those names are your inside track for discovering the contact information of a specific researcher!) Spend a few minutes reading a poster to determine whether the research appeals to you. If it does, write down three specific reasons, along with the names and contact information of the first and last authors listed. (*Note: Don't take a photo if you don't have permission.* We know this seems odd, but some researchers have good reasons to be picky about this—yes, even if they choose to display their poster in a public setting.) Then, upload the contact names and poster title to your reference manager or to the folder in a cloud storage app with an appropriate label such as "Potential Research Opportunities." You'll return to this folder when you're ready to apply for research positions.

ATTEND A RESEARCH SYMPOSIUM ON CAMPUS

When it works perfectly, this is an efficient method to conduct your search and interview at the same time. But even when the process isn't so streamlined, on-campus symposia are ideal opportunities to learn about the research happening at your institution and meet a potential mentor—or several. However, one potential drawback is that a symposium might not be held when you're ready to search for a research position, so simply because of timing, you might not be able to use this approach.

A research symposium might be organized by a department, research institute, student research group, or campus organization. Sometimes the only presenters are undergrads who belong to the same department, but in other cases multiple departments cohost the event. Some symposia are open to any students who wish to present their research, whereas participation for others is a requirement for students enrolled in research for course credit that counts in their GPA. Although each organizer determines the format of their symposium, the focus might be on poster sessions or be a mixture of short research talks (possibly given by undergrads, postdocs, or grad students) and research posters on a variety of topics.

If it works with your schedule, attend the poster session of a symposium. (For details on when it's scheduled, email the organizer by contacting the department or other entity that is hosting the event or check their Facebook page.) A poster session is essentially when researchers hopefully wait by their posters to explain their project to anyone who wanders by. The advantage for you is that the person who conducted the study is typically available (and happy) to give a tutorial on their project. And if you like what you learn, you might turn the interaction into an interview for a research position working directly with the presenter or another member of their research group.

At the session, when a title of a research poster catches your attention, take a few minutes to read the abstract or introductory paragraph. If you're at the poster at the same time as the presenter, pay close attention to the conversation they have with other attendees about their research. When the opportunity arises, ask a question to the presenter, but adjust it depending on who that person is.

If, for example, the presenter is another undergrad, ask, "What is the overall question you're trying to answer?" The question might be the same as the poster title, or it might not be. If the research topic still holds your interest, ask for their mentor's name, title, and email address. You could also ask the undergrad presenter to introduce you to their research mentor if that person is at the event. Finally, ask the undergrad presenter who their PI is and for their PI's email address.[2] But if the presenter doesn't have contact information for either their mentor or the PI, you can get the information from the campus directory. (You'll learn how to use what you've learned during this interaction to send an impactful email in chapter 5.)

If there is time, you can also ask the undergrad presenter some general questions about their research experience that relate to the lab's culture and their mentor's training approach. We've included a section on maximizing this type of interaction in a later subsection titled "Ask Your Classmates about Their Research Experiences and Evaluate Their Answers," but you might only have time for one or two questions at a presentation. Before moving on to review the next poster, remember to thank the presenter for their time—after all, you might become one of their labmates.

However, if instead of an undergrad researcher, the presenter is a grad student or a member of the professional research staff, their presentation style might be intended for those with advanced knowledge

2. A note about posters and author names: more than one PI can be listed as an author on a single poster. You'll want to know which PI the presenter works with if multiple names are listed. Also, although typically the first author listed on the poster is the person presenting the research, confirm this to be the case. You'll need to know the correct contact person when you email a PI or potential mentor.

of the field. Such as presentation can be frustrating and confusing for a nonexpert, but ultimately it won't prevent you from leaving with what you came for. When you have the opportunity to ask about their research project, **choose a broad question that will help them shift gears into more accessible language.** Good questions might be "What is the big question you're trying to answer with your research project?" or "What is the specific question you're trying to answer right now?" followed by, "Why is that important?" You don't need to gain an in-depth understanding of their research at this moment—you just need to learn *something* and show genuine interest.

After you've absorbed their research synopsis or response to the question you asked, if you'd like to conduct similar research or assist with theirs, mention that you're an undergrad in search of a research project and that you attended the symposium specifically to learn about the research on your campus. State that you're interested in their project because ___, and fill in the blank with something specific that they told you or the reason the poster originally caught your attention. Next, ask, "Will you consider me for an undergrad research position? I have *X* hours per week to dedicate to research this semester." Yes, this can feel awkward, but you don't want to hint—you want to ask for a research position politely, directly, and with the most confidence you can muster. The ability to do this, no matter how uncomfortable it feels, will only be to your advantage.

The next move is the presenter's. There are several possibilities that might happen, but some are that they might (1) conduct a mini-interview with you on the spot; (2) set up an interview appointment; (3) request that you email them with specific documentation, such as a resume or transcript; (4) refer you to their PI; (5) state that they don't have the time to train an undergrad researcher and refer you to someone else in their research program; (6) set up a time for you to observe them in the lab; or (7) state that their lab doesn't mentor undergrads or does not have an open position right now. If the presenter's response is any of the responses 1–6, you've identified a potential research experience and possibly a research mentor. Before you move on to the next poster, even if they gave reason 5, confirm that the presenter is the first author listed on the poster and which author is their PI.

If the presenter's response was any of the responses 1–5, follow up with them or the labmate they suggested to you by email within twenty-four hours. Don't wait any longer, because you want to take full advantage of the groundwork you laid by attending the symposium. Also, sometimes researchers are excited to start mentoring a new student, but if they have a couple of weeks to think about it, they might lose that motivation and change their mind. Plus, the sooner you follow up, the more enthusiasm and genuine interest in the research you demonstrate. (The chapter 5 section "Step 2: Make First Contact" includes tips on how to make the best impression when you follow up.) We summarize this section by presenting nine tips for attending a poster session.

Nine Tips for Attending a Poster Session at a Symposium

1. **Register when required.** Some symposia have a registration deadline for attendees who are presenting posters, giving a short talk, or planning to attend a reception, even if there is no fee to attend. Also, registering might give you access to all the poster abstracts, which can help you decide ahead of time which posters to prioritize. Not all symposia will have printed abstract books, but many make the full text available online. Some symposia don't require registration to attend, in which case you can just show up and learn. Whether you need to register, and other questions, will be covered on a symposium's website or, if it's a department or student-managed event, on a related Facebook page.
2. **Start with realistic expectations.** Even at a poster session with an expert explaining their project or overall research program, some information will be tough to grasp. Plus, you might not yet have the scientific background to gain a thorough understanding of the research at a poster session—and that's okay. In addition, everyone who is an expert in their project isn't necessarily an expert at explaining it. Keep this in mind if you become overwhelmed or start to feel unintelligent during any research presentation.
3. **Follow basic phone etiquette.** Put your cell phone on silent or vibrate and keep it in your pocket or backpack except when you take photos or notes. Essentially, don't text a friend or check your social media feed while someone is giving a presentation in front of you.
4. **Take notes.** Jot down a few notes on why a research study is important or why it inspires you. You can do this while someone is presenting their poster or after you've asked them a question (these two are our top recommendations). If you ask a presenter a question, write down their answers because you'll use their answers later when you contact their PI or another research group member. If it's not possible to take notes during a poster presentation, do so prior to visiting the next poster because the presentations to blend together—even at small events.
5. **Ask for permission before you take photos.** If asked, many, but not all, presenters will allow you to take a photo, but be polite regardless. *For a variety of reasons, some presenters won't want the prepublication data on their poster photographed.* If a poster is unattended (as odd as that might seem), it's acceptable to write down the title and poster authors and some information about the research, but don't take a photo just because there's no one there to ask for permission.

6. **When permission is granted, take high-quality photos.** If you're given permission to take a photo of a poster, make sure the entire poster is in focus or take several shots. You'll also need the contact information to follow up with the lead author or the PI. If you're given permission to take a photo, ensure that the entire poster is in the photo before uploading it to your reference manager.
7. **Don't waste a presenter's time.** Not all posters will spark your interest. It's acceptable to read the title, perhaps some text, and move on if you're not interested in learning more. You're not required to hang out while someone gives a presentation on a topic you don't want to learn about or to ask fake questions because you're afraid of hurting a presenter's feelings. Researchers know that not every person at a symposium is fascinated with their work. Giving a polite smile and nod to the presenter is all you need to do as you pass by. Your goal at an on-campus symposium or meeting is to make connections with researchers working on topics that you're genuinely curious about—so make sure to spend your time doing that.
8. **Don't let an unenthusiastic presenter ruin your experience.** Presenting can be tiring, and it's possible that you'll want to learn about a project but a presenter won't be interested in discussing their poster. Unfortunately, sometimes a presenter who has reached their limit of interacting with attendees doesn't express it well. Although it's possibly rudeness on the part of the presenter, it's not a value judgment against you or your academic status as an undergrad, and you must not let their behavior discourage you. Just move on to the next poster.
9. **Be courteous to others.** Venues can be crowded, and sometimes posters and presenters are packed into a very small space. If you want to take a break from the poster session to check your phone, text a friend, or have a social conversation with someone, make sure to move far enough away from the posters that you don't block the other attendees' access.

ATTEND A RESEARCH FAIR ON CAMPUS

Through organized events, often referred to as *research fairs*, some campuses connect mentors searching for an undergrad researcher with undergrads seeking a research opportunity. Although the formats vary, research groups and various administrators might set up tables with information on undergrad research, open opportunities, and campus wellness resources. Similar events with a less formal vibe include those at which PIs or group members wander around the same room with undergrads while eating light snacks. These "meet and greet" opportunities allow undergrads to ask questions about research in general, inquire about career path options, and discuss available research opportunities with specific mentors. At most research fairs, the atmosphere is casual, but you should still approach each interaction with a PI, group member, or administrator as a mini-interview—for both you and the potential mentor.

ATTEND A SEMINAR THAT SEEMS INTERESTING

Departments, research groups, and research institutes often host seminars throughout the semester. Some do so as part of a regular seminar series with weekly, biweekly, or monthly talks, and others do so as an occasional specialty series. The seminar presenter might be a PI from your home institution (or one visiting from elsewhere), a postdoc sharing their latest results, or a grad student defending their thesis.

Attending a seminar can inspire you to work on a particular project or research topic and help you to both identify and impress a potential mentor. Seminars are generally open to everyone. So, if you learn about one from a professor or notice one advertised on a bulletin board, a departmental web page, or social media page, or you receive a notification about one from a Listserv, you're invited to attend. You can attend seminars in person or possibly online. Some seminars are also recorded. To find an archived talk, consult the same source where you learned about the event. Ideally, before you attend the seminar, spend some time on the internet reading the presenter's research interests and familiarizing yourself with the research topic. At the very least, do a web search using the entire title of the seminar as the search term.

Also, although you could go to doctoral and master's degree defenses, we don't recommend attending as a strategy to connect to persuade the presenter to be your research mentor. A grad student who is defending their thesis or dissertation won't be available to mentor a new undergrad researcher.

What to Do When You Arrive at a Seminar

First, find the posted seminar notice. Take a photo and upload it to your reference manager or file it in a cloud storage app that you've labeled "Potential Research Opportunities." Make sure you have the seminar title, date, name of the presenter, and the host's name (if applicable) in the photo. Even if you previously snapped a photo of the notice from a bulletin board, do it again because sometimes the seminar title changes. The seminar title must be accurate when you email a potential mentor. If there isn't a seminar notice posted outside the room, write down the date, presenter's name, and title of the talk when the presenter is introduced to the audience. If the presenter is from your campus, you'll contact them about a research position. But also write down the name and department affiliation of the person who introduces the presenter—this person is usually hosting the seminar speaker. If the host is from your campus but the presenter is not, you'll contact the host to ask for a research position.

You Have One Job: Leave with What You Came For

During the seminar, make it your goal to write down three reasons, phrases, or sentences that highlight why the presenter's research is interesting. Although there might be several topics to choose from, some could be a result, the overall study, a statement about the project background or significance, the future directions the researcher plans to study, or a technique you're familiar with from a class. The more specific your notes are, the bigger the advantage you'll have when you apply for a research position.

Give Your Full Attention to the Presenter

When you attend a seminar, you'll want to elevate yourself to the same professional level as the other attendees (or above, if you're by someone who is rude). A seminar room is filled with busy professionals who believe the presenter has something to contribute either to their own research or to the world at large. Often, a seminar might be the only chance the researchers in the audience will have to attend a talk given by the presenter. Most attendees will be professors, professional research staff, and grad students. **Although undergrads are welcome to attend seminars, unless a professor is awarding extra credit for doing so, it's less common. Therefore, undergrads who do attend have a greater chance of being noticed.**

We'll share a story that underscores the importance of this very point. One of our colleagues occasionally forwards seminar notices to students who express an interest in joining their research group. Our colleague finds that almost no students attend, but those who do either become inspired by the research topic or realize that it's not something they wish to pursue. One semester, our colleague informed several students who inquired about joining their research team about a seminar closely related to their research program. On the day of the seminar, although the room was large, it was packed with some attendees standing in the back or sitting on the auditorium stairs. Approximately fifteen minutes into the seminar, an undergrad student a few rows away from our colleague proceeded to remove several items from their backpack and place them on the desk: a binder of class notes, a highlighter, a pen, a bag of candy, and hand sanitizer. The student then began what our colleague presumed was the student's typical routine in a class lecture: they applied the hand sanitizer and spent the rest of the lecture highlighting notes in their binder while pausing occasionally for a very crunchy treat from their snack selection. On occasion, the student would stop to check their phone, presumably to send a text or check a social media account or glance at the presenter in the front of the room.

Beyond this ritual creating a distraction for our colleague and everyone else around them, the student took a place that could have been occupied by someone who *was* interested in the seminar. Imagine how short the research interview was when that same student showed up a few days later at our colleague's office in hopes of securing a research position. The student had no hope of convincing our colleague that the student was interested in our colleague's research. Although this is perhaps an extreme example, even if the student had spent the seminar distracted by their phone instead of the panoply before them, our colleague said that the interview would have been just as short. As an undergrad, you may find that your attendance might be noticed—it's up to you to make it an advantage in your search, or at least not a disadvantage at an interview.

Four Tips on Seminar Etiquette That Can Make All the Difference

1. **Sit in the back in case it's difficult to focus (or stay awake).** If needed, distract yourself by taking notes on a legal pad or discreetly doodling in a notebook to keep yourself from doing the falling-asleep head bob.
2. **Turn off your cell phone.** In a seminar, it's bad form to text, check social media, or otherwise broadcast that you have something more important to do than give the presenter your attention. In a professional seminar, this matters. *Even if everyone else around you is live tweeting the seminar, stay off your phone*. As an undergrad, what you do will be perceived differently than what a professional researcher does. It's not fair, but it's true. After you've joined a research group, if you want to participate in science communication by live tweeting a seminar, go for it—because your labmates will know that's what you're doing, and the opinions of other attendees won't matter.
3. **Take notes on paper.** Again, use a legal pad or a paper notebook because some people believe that anyone who is "taking notes" on an electronic device is actually on social media or playing games. This isn't just an unfair poke at undergrads. There are faculty members who believe that no one in a seminar could possibly be taking notes and are openly critical about it. Simply take notes on paper to avoid this misunderstanding.
4. **Don't leave the seminar early.** If you don't have enough time before your next appointment to stay for the entire seminar, then skip the presentation. Most seminars last around fifty minutes—you can get through the talk even if you're bored. You can respectfully leave when, at the end of the lecture, the presenter invites questions. However, if you have the

time, stick around for the question-and-answer session because you might learn something. Sometimes the Q and A portion is the best part of a seminar—in particular, when an audience member with limited background knowledge in the subject asks the presenter for clarification.

A Note on Why It's a Bad Idea to Claim You Attended a Seminar If You Did Not

Seminars get canceled. They get rescheduled. A talk title can change at the last minute, and on occasion so can the presenter. Sometimes seminar rooms are small or only a handful of people attend. It's not worth the risk to falsely claim attendance at a seminar that you didn't attend. You'll risk eliminating yourself from a position due to a "character issue," and once a mentor has decided to pass on a student because they suspect this might be the case, there is virtually no way to be reconsidered for that research position.

ASK YOUR CURRENT LECTURE PROFESSORS WHETHER THEY HAVE AN OPEN POSITION IN THEIR RESEARCH GROUP

In the sciences, many professors direct research programs in addition to teaching classes. Typically, class subjects overlap with a research specialty, so if you're interested in the subject a professor teaches, you also might be interested in their research program. Some professors even recruit students from the classes or labs they teach. Strategies on connecting with your lecture professors are covered in chapter 5.

ASK YOUR GRADUATE TEACHING ASSISTANTS (AND LET THEM RECRUIT YOU)

If your lab or lecture course is taught by a graduate teaching assistant (TA) in the sciences, it's likely that they also conduct research. Although that research might or might not be in the same field as the lab class they teach, you might be genuinely interested in what they study. Your TA probably has an online profile on their department's website that covers their research interests and current project(s). By reading this research description, you should be able to determine whether their project is something that you'd like to be part of or at least would like to discuss the possibility with them.

After you've read up on their project, but before you ask for a research position, we highly recommend that you read and then craft an Impact Statement. (Details on how to do this are presented in chapter 5, in the section "Step 2: Make First Contact," subsection "The Email Template.") Then, **approximately three weeks into the lab or lecture course, or earlier if you're prepared, approach your TA and state, "I read about your research project online and would like to learn more about your project because ____" (fill in the blank with your Impact Statement).** Then continue with, "I have *X* hours per week to dedicate to research. Will you consider me for a research position?"[3]

What you do next depends on their response. If your TA says that they don't have an available position, ask them if they know of anyone who does. In person, to a TA, you can be broad when sharing information about your research interests, and it probably won't count against you. Plus, compared with faculty members, grad students are more likely to have a conversation about unfilled undergrad positions with the members of their cohort. If your TA doesn't say yes but doesn't say no, it's not a bad sign. If they are interested in mentoring an undergrad researcher, they will likely set aside time to interview you later—and you can certainly ask them for an interview if they don't offer one. Alternatively, they might refer you to someone in their research group or to a listing of available undergrad research projects posted by a departmental administrator. They also might not take any immediate action, which essentially puts you on a semester-long interview. This can be a huge advantage if you perform well or a disadvantage if you're routinely unprepared for class or rude to your classmates.

At the end of the semester, or possibly sooner, you might find that the TA recruits you. If your TA doesn't follow up with you within four weeks of your asking about a research position, connect with them again about the possibility of joining their lab. If their response is still noncommittal, direct your search elsewhere. They might not know how to politely tell you that they aren't interested, for whatever reason. Or it they might be unsure whether they have the time to mentor an undergrad in research and are still mulling it over. If either of these are the case, try not to take it personally even if it feels so.

Six Tips to Demonstrate to Your TA That You'll Be a Good Labmate

By implementing the following tips, you'll show your TA that you take your class responsibilities and lab safety seriously, handle disappointment as

3. If your class is taught by an undergrad TA, it's worth asking if they also conduct research in a professional lab. Although it's unlikely that an undergrad TA will be able to offer you a research position, asking about their experience could be enlightening. And if you ask directly, they might agree to connect you with their in-lab mentor, which could lead to an interview.

an opportunity to solve problems, and are comfortable working independently as well as being a supportive group member. And if your course is taught by a professor and not a TA, all the following tips still apply—only you'll ask the professor for a research position.

1. **Let your approach to the instructional lab model how you'll approach your responsibilities in a professional research lab.** Show up on time to lab class (don't keep your lab partners waiting or miss your TA's initial announcements) and ensure that you've prepared ahead of time (read the prelab, if required). Complete all extra-credit assignments that your schedule allows time for and put focused attention into completing all lab assignments (don't ask a lab partner to cover for you so you can leave class early), and follow all instructions for lab reports and papers. Turn in all assignments on time or connect with the TA as soon you know that you'll need extra time to complete one.
2. **Participate but don't dominate.** Depending on the lab class, you might complete the assignments by yourself or you might choose, or be assigned, one or more lab partners. If there are group assignments, always contribute your fair share to the effort but don't try to control the other members. Your goal is to show that you're flexible and can work as a cohesive team member but are also comfortable taking the lead on occasion. (Plus, practicing this in an instructional lab will be helpful when you join a professional lab because you'll need to find this balance with those labmates as well.)
3. **Demonstrate perseverance.** If a module is challenging or fails, keep a positive attitude during the lab and on class-associated learning management systems. (TAs often, and professors sometimes, read the comments on forums.) Don't complain that your lab partner is to blame or that the class is disorganized or a disaster. If you give the impression that you have no patience when a procedure doesn't immediately work, you'll send the message that mentoring you in a research project might be unproductive for both you and an in-lab mentor. When a lab class procedure fails, or an assignment doesn't go smoothly for whatever reason, avoid giving in to frustration or complaining. Instead, follow the strategy described in the next point.
4. **Show that you want to solve problems.** If you, or a lab partner, makes an error, try to approach your TA with the question, "What can I or we do to fix this mistake?" or with an idea to that might solve the problem. But even if you have no idea what your options are, that's okay. The point is to demonstrate that you want to solve the problem and move on—not complain about the situation or blame the TA for designing a subpar module. Then, even if it's disappointing because you'll need to redo a procedural step or

throw everything out and borrow results from another lab group to write your lab report (with your TA's permission), just chalk it up to it being an undesirable outcome and get the task done. This is an opportunity to show that you can manage such annoyances, which will probably be part of any research experience.

5. **Be patient with lab partners.** As important as it is to remain patient when a lab procedure is finicky, it's also essential to demonstrate that you're considerate to your fellow group members if someone makes a mistake, a lab partner is confused during the procedure, or a group mate isn't as skillful as you are when completing a task. Granted, it's frustrating if a partner arrives unprepared for class or doesn't read the instructions carefully and the result is a compromised or ruined experiment or procedure. But if any of these things happen, remind yourself that although it's frustrating, it's not tragic. (If such a situation affects your grade, however, make an appointment with the TA to discuss it to find a solution.) Your interactions with your classmates will indicate how easy (or difficult) it will be to work with you in a professional research lab.
6. **Demonstrate self-reliance.** You should ask for help or clarification when you need it but first ensure that the answer you seek isn't already in a protocol, the lab manual, or the syllabus before checking with your TA. Although not all your questions will be answered in these documents, your TA and professors put a lot of effort into covering the most commonly asked questions in them. For example, if your question is about when something is due, how long a lab report should be, or what it should cover, or on a similar topic, check the syllabus—it will have the answers. In the lab, carefully read the protocol before asking for clarification. You'll make a negative impression if you continually ask your TA to essentially read instructions to you. Most researchers want the undergrads they mentor to become semi- or fully independent in their research experience, and showing that you're as self-reliant as possible in a lab class is a good indication that you'll rise to the challenge in a professional environment.

CONSIDER RECONNECTING WITH FORMER TAS AND PROFESSORS

If you've already completed a lab class, consider contacting any former TA or professor who would remember you from that semester. Usually, the current profiles of grad students are listed on their department's website and include a link to a description of their thesis project and why it's important. Their email address is typically included in the profile, but if not, use the campus directory to find it. To determine whether you'd like to work on a project with a former professor, use the strategy on exploring faculty research interests previously described. Then, as you would with a professor, contact them through email or attend their current office hours.

CHECK WITH STUDENT ORGANIZATIONS

Start by examining the web pages (official college or university pages and social media) and Twitter feed of student organizations on your campus. Consider preprofessional societies, honor societies, and groups for a specific major or career path. **Basically, if the group's members participate in undergrad research, it's worth exploring.** The group might post advertisements for PIs or mentors searching for undergrad researchers. These open positions will likely fill quickly, but it's an easy option to explore.

If the group hosts seminars from researchers on your campus who are searching for undergrad researchers, attend these meetings. **Before a talk begins, snap a photo of the slide that has the presenter's name, title, and department or center affiliation.** Whether or not you talk with the presenter, you'll want their contact information so you can email them shortly after their presentation. **During the talk, write down three specific statements or phrases about their research.** It can be a technique such as "fluorescent labeling of proteins in living cells" or the central question addressed by the presenter's research, "Our lab studies a family of proteins that inhibit . . . ," or a specific result, "We found that protein A interacts with protein B . . ."

If you have the opportunity to discuss a potential research experience with the presenter after their talk, remember to approach your conversation as if it's an interview, *because the presenter definitely will.* If you would rather follow up with them by email, do it within a few hours and no later than twenty-four hours. The sooner you follow up, the more you demonstrate your enthusiasm and more likely you'll beat your competitors to the inbox. Chapter 5 explains how do to this well.

ASK YOUR CLASSMATES ABOUT THEIR RESEARCH EXPERIENCES AND EVALUATE THEIR ANSWERS

By asking a few questions, you'll be able to determine whether a classmate's research experience seems like something you'd enjoy. However, **remember that whether they like or dislike their research experience isn't as important as the *reasons they feel the way they do.*** For instance, if they like their research experience because they get plenty of time off to study for exams, or regularly skip when there isn't anything to do, would that experience be a good match for you? What if they dislike the experience because they spend most of their time observing? Would you form the same opinion if you had a similar experience? If they feel overwhelmed by the weekly hour commitment, is it because their mentor's expectations are too high, or they haven't found an academic and life balance that works?

It's not enough to ask your classmates about their research experiences; you need to evaluate what they tell you based on your goals and the expectations you have for your research experience. And you need to remember that your classmates will be passionate (or not) about their experience because of their values, what activities they give up to participate in research, and their goals for a research experience. Also keep in mind that they currently might be frustrated with a difficult technique or a part of their project, which could negatively influence how they feel about their research experience in the moment. Perhaps in a week, they will be through the rough spot and once again be inspired.

Questions to Ask Your Classmates

For the conversation to be the most effective, ask for specific information about their research experience. Here is a list of the most important questions you should ask.

- What do you like the most about your research experience?
- What do you dislike the most?
- What has been the most challenging part of your research experience so far?
- What is the name and title of the person you initially contacted about the position? (If they applied to a specific person, contact that person if you decide to pursue an opportunity in the same lab. If they don't want to give you contact information, you can find it in the campus directory.)
- How many hours per week do you spend working on your research project, and in what blocks of time?
- Do you set your schedule, or is it set by your mentor?
- What techniques have you learned?
- What do you spend most of your time doing? Or what is a typical day in the lab like?

The following question is not essential but will provide additional information:

- What are the objectives (or specific aims) of your project?

If your classmate applied in response to an advertisement, ask where it was posted or how they became aware of it. Check the same place for a current advertisement. If there isn't one, don't be discouraged. If you're interested in becoming a member of the research group, follow up with the person they applied to, or contact their PI.

SIGN UP FOR EVEN MORE EMAIL (REALLY)

Join all email Listservs related to your major, preprofessional groups, and clubs that are research related. **Skim each email the day it arrives.** Pay attention to interesting seminar announcements, research symposia, or available research positions. **If you read an announcement for an undergrad research position,** know that the professor or mentor will be inundated with responses within a few hours. So, you'll want to **respond as soon as possible, with a custom Impact Statement and any supplemental materials asked for in the email** (transcript, CV, or other). This is one of those cases where the first few responses might be the only ones read.

CHECK OUT PIS' TWITTER ACCOUNTS

Increasingly, PIs are using Twitter to announce vacancies in their lab team. Although most usually recruit professional researchers and grad students on social media, some also share posts about open undergrad positions. If you're interested in working with a particular PI (or other member of their research team), do an internet search with "[their name] + Twitter" and you might find their account. Other PIs or group members include their social media channels on their lab websites. Although more of a long shot, it can be worth a quick check to determine whether they have tweeted about any undergrad research opportunities in the last few weeks. If you don't find such a tweet, you can @ them or send a direct message to ask if they have any positions available. However, before you do, make sure your Twitter feed is completely professional or start a new account. And even though Twitter is an informal venue to connect with someone, follow the guidelines in the chapter 5 section "Step 2: Make First Contact" even when connecting with a potential mentor on social media.

EXPLORE OUTSIDE YOUR MAJOR AND YOUR COLLEGE

Regardless of your major, consider searching outside your home department for research opportunities. Many research programs use a multidisciplinary approach and train students from various majors. For instance, undergrads in our lab have majored in chemistry, physics, biology, molecular biology, microbiology, genetics, bioengineering, psychology, anthropology, and biochemistry. The majority have been premed (or prehealth), but others have been predental, prevet, pregrad, or planning to enter the job market after graduation. Your aim should be to participate in a research project that excites you and presents opportunities to accomplish your personal, professional, and academic goals. Whether or not the research experience is in the same department as your major isn't always a relevant factor.

TIPS ON SELECTING AND READING A SCIENTIFIC PAPER BEFORE APPLYING FOR A RESEARCH POSITION

To identify potential research groups or projects, often undergrads are encouraged to read a few of a PI's *publications*, also referred to as *peer-reviewed journal articles*, *primary literature*, *articles*, or *papers*. In theory, this is a straightforward task. In practice, many students find the activity overwhelming—especially first-year undergrads or those who haven't completed much classwork in the research field where the paper was published. Unfortunately, it's easy for students to misinterpret their struggle to comprehend a journal article as an intellectual failure—which it isn't—and self-disqualify themselves from pursuing an undergrad research opportunity. This seems to be particularly true for those who presume that all their classmates were successful with the task because they found research positions.

But many mentors (including us) don't require undergrads to read a peer-reviewed article until they start conducting research. And when we asked colleagues at multiple career stages and in multiple scientific disciplines for their policy, even if they recommend doing so, most felt that it was unnecessary as long the student demonstrated self-directed learning about their research program *in some way*, although we do need to mention that a few shared that reading assigned articles is part of the interview process for their group.

In any case, **the hidden curriculum here isn't the advice to read a paper but the knowledge that the task isn't always *required*.** However, if you mention that you've read one of the PI's papers (or one is sent to you from a potential research mentor as part of the application process), it's reasonable to expect that you'll discuss the article in some way at an interview and likely share what you learned or found noteworthy about it.

But if you didn't read an article, don't claim that you did: it will be obvious within seconds that you didn't because you can't fake your way through a discussion of a scientific paper. Although we don't expect undergrads to read journal articles prior to joining our research program, during interviews several students have volunteered, "I read a paper of yours," as if stating this was a checkmark in the interview process. However, when asked the inevitable follow-up questions, "Which one?" or "What was the most interesting part of the paper?" the answers have always been a flustered, "I don't remember," or a variation on that theme. Typically, the next moment has been awkward as the student quickly states that maybe it wasn't a paper, or they tried to read a paper, or that they are confused and didn't mean that they read a paper, or that they *planned* to read a paper. In any case, it's disappointing—not that the student didn't read the paper but that they misrepresented doing so. Colleagues who have had similar experiences echoed the same disappointment.

Reading the current scientific literature is challenging when you're starting out and especially so when don't have a mentor to guide you through the process. However, this doesn't mean that you shouldn't try to do it. It simply means that if reading through a paper becomes frustrating, don't let it derail your search. But before attempting to read a peer-reviewed journal article, we recommend first reading an article on the best practices for, well, learning to read a scientific paper. There are numerous free blog posts and articles with strategies that you'll find by doing an internet search with the terms "how to read a scientific paper." After you've read a couple of those, if you do choose to read a paper, or if the PI or your potential in-lab mentor gives one to you as part of the application process, keep the following tips in mind.

TEN TIPS FOR SELECTING AND READING SCIENTIFIC PAPERS

1. **Find some papers.** To find a published journal article, search PubMed or Google Scholar using the PI's name or visit their website to view an abbreviated list of their publications. (This might or might not be an updated list.) Choose two or three articles at the most but focus on reading and understanding a single paper in the beginning. Move to the next paper if you get stuck on the first one.
2. **Add each journal article to your reference manager before you start reading.** The best time to highlight and annotate a research paper is while you're reading it. So, add each article to your reference manager after identifying it. This is an easy habit to develop, and if you establish it before you start undergrad research, you'll be grateful when reading articles becomes an expected task. And if you continue with grad or med school or in a science job where reading articles is a given, this habit will be even more important.
3. **Don't give in to frustration.** For papers in some disciplines, you won't understand the title without performing an internet search on several words, much less the results, significance, or conclusions. But keep trying

even if you find most of the language or concepts confusing. You might not have the background knowledge or scientific vocabulary to gain a meaningful understanding of what you read, but you can still learn something. *It's important to remind yourself that difficulty is to be expected and not let it throw off your enthusiasm.*

4. **Set realistic goals.** Your goal isn't to become an expert in the subject or to be able to discuss the paper at length but rather to introduce yourself to a topic that is important to the PI and learn what you can. If, while reading a paper, you determine that the research study is inspiring, consider it a bonus.
5. **Use resources to decipher the terms.** Use web searches and textbooks to decode the significant terms. Find the definitions of short terms, such as "epigenetics," or longer ones, such as "laser scanning confocal microscopy." You might find it easier to digest the information this way, even if it's generalized. (A side benefit is that the additional search results will help expand your overall knowledge base; for example, even if you don't end up approaching the PI who wrote this paper, perhaps epigenetics will turn out to be a field that excites you.)
6. **Focus on the summary section (or introduction).** Once again, use a web search to glean the overall focus of the paper and decide whether it's interesting to you. After you read the first few sentences of the introduction or abstract, you might understand the general problem that the paper addresses.
7. **Skim the methods and materials section.** Determine whether any techniques are ones you learned about in lecture or performed in an instructional lab class. If so, you'll have a point of reference. Again, do a web search to read about or, when available, stream videos of specific techniques. At this stage, try to learn why the techniques are important to the research study—not just how they are done.
8. **Try to determine authors' roles in the research.** Papers are often collaborative. A paper might list multiple authors from a single research group or be a collaborative project that was completed by members of multiple research groups. Therefore, a technique or experiment you find interesting might not have been done in the lab that you're hoping to join. Some papers denote each author's specific contribution, but many don't. Check the author contributions section of the article; it should state who did what work. Be aware of this if you plan to contact a PI based on a specific technique or an experiment described in a paper.
9. **Be aware that some papers are published on projects that have been completed.** These projects are typically related to the overall research focus of the lab in some way but are no longer being pursued. Even if a specific project is no longer available, if you like the research focus of the lab, it's likely that other, similar projects will be available.

10. **Revisit the paper after you join a research experience.** After you've begun research, you'll have someone (or possibility several people) available to guide you through the process of reading a journal article. You might be assigned papers by your mentor or you might want to revisit ones you identified during your search for a research experience. For some students, this will be expected during their first semester.

When you apply for a research position, you'll always be better off being honest about what you have read—even if it's nothing.

MAINTAINING YOUR ACADEMIC AND LIFE BALANCE IN FUTURE SEMESTERS

Every semester is a do-over with respect to scheduling your academic and life balance. Therefore, continue the strategy presented earlier in this chapter to schedule classes, activities, and research hours. **Evaluate your schedule often and add, reduce, or eliminate items as needed to avoid becoming overextended.** The small effort spent scheduling is worth the big payoff in productivity, stress reduction, and overall happiness. Plus, the self-discipline and time management skills you'll gain by periodically evaluating your schedule throughout college will help you develop the skills to do this for the rest of your life.

But keep in mind that maintaining an academic and life balance is so much easier said than done. And we assure you that it's not just a struggle for undergrads—after all, there is an entire industry dedicated to time management strategies aimed at professionals who "should" have it figured out. We even wrote an entire book on this subject, *Life and Research*, to help early-career researchers navigate this complicated process.[4] But for you as a student, at the very least, it's imperative to set your priorities and learn to recognize the signs of overcommitment.

When you overcommit your time and get behind in the social or professional obligations you've made, it adds stress to your life—sometimes a lot of stress. You end up being stressed about the studying you're not getting done, stressed about letting your friends or family members down, and stressed about the obligations you abandon just to keep your head above water. All this stress takes a toll on your capacity to learn and recall infor-

4. Paris H. Grey and David G. Oppenheimer, *Life and Research: A Survival Guide for Early-Career Biomedical Scientists* (Chicago: University of Chicago Press, 2022). Although this book is primarily intended for grad students, postdocs, and staff scientists, much of the advice is applicable to undergrads in postbac positions or students who are in their final year of undergrad and are preparing to attend grad school.

mation. If it gets bad enough, you become unable to focus during lecture classes or while reading assignments, so you retain less information and end up being even more stressed as exams approach. So inevitably, you try to solve the problem in the short term by cutting class, getting less sleep, reducing your research hours, cramming for exams, or doing subpar work on your out-of-class assignments. This in turn could compromise your academics, recommendation letters, happiness, and potentially your health. Worse than the feeling of letting a club or a friend down is the personal toll overcommitment takes on you, because you set yourself up for failure instead of success.

Every part of your life is more difficult when you're overcommitted. Make a conscious decision to evaluate your schedule before adding new activities, or quickly reevaluate and prioritize if you start getting in over your head. And be honest with yourself about where your time goes, or no amount of thinking about your schedule will help. For example, if two hours of scheduled personal time regularly turns into five, or one hour of scheduled study time often turns into fifteen minutes of study time and forty-five minutes of personal time (or nap time), don't waste energy feeling guilty about it—just make changes.

Sticking to a well-managed schedule might not be enough to maintain the academic and life balance that supports your needs and well-being. If you're having difficulty handling the stress of college or life in general, for example, or difficulty organizing your time; passing exams or classes; managing a disability or a chronic illness; dealing with hostile or inappropriate behavior directed at you because of your race, ethnicity, gender, or gender identity; or struggling with any other issue, then make an appointment with the appropriate office(s) on campus to ask for advice. These might include visiting the Dean of Students Office, Ombuds Office, Academic Advising Center, Student Counseling Services, Office for Accessibility and Gender Equity (or the Title IX coordinator), or another office. A growing number of campuses also maintain open pantries where students who are food insecure can pick up free items without having to show proof of need.

To find the contact information of these and other resources, you could do an internet search such as "student academic support [your college] or "student disability center [your college]" or "student support resources [your college]." Using these search terms should return a listing of multiple resources that range from financial and academic assistance to LGBTQIA+ and crisis services. Depending on your campus, you might have the option to make and attend appointments either virtually or in person. You can also go to office hours of any of your professors or TAs to ask for tips or connect you with one of the previously mentioned resources or one they recommend.

It's unfortunate that well-being discussions aren't always at the forefront of college life, but they should be. Asking for assistance when you need it isn't a weakness or something to be ashamed about. It's not a failure to ask others for advice or for help in any area of your life. We know that asking others for help might seem awkward at first—especially if you want to figure out everything for yourself or you've been misinformed that asking for assistance signals that you're not ready to be in college. But **you'll get a lot farther in college (and in life) if you're able to embrace the idea that you can learn from others' experiences and expertise if you're willing to ask for help.** The key is that you find someone (or a few people) who will help you learn to manage your stress or help in other ways as needed. And don't believe that no one you know is using these resources. They just aren't talking about it.

5
Your Application Strategy

Sometimes the application process for an undergrad research position is a multistage, formal undertaking that involves filling out an online form, writing essays, uploading academic transcripts, and, if you make it to the next level, meeting for an interview. In other instances, the process might be as simple as asking a principal investigator (PI) for a research position during office hours and discussing a project and the PI's expectations right then and there. Many application processes land somewhere in between. You approach a PI or other potential mentor, share a genuine and enthusiastic reason that you're interested in joining their research group or project, and secure an interview.

Regardless, when you contact a potential mentor to ask to join their research group, you're applying for a research position. For many potential mentors, their first impression of an undergrad is the one they use to determine whether they should offer an interview. Therefore, whether in person or through email, to be the most competitive candidate when you connect with someone to inquire about doing research with them, it's essential to apply the same strategies that those with knowledge of the hidden curriculum do.

The list that follows presents the ten most common mistakes undergrads make when applying for a research position, tips on how to avoid those mistakes, and why you should. In the interest of brevity, for the rest of this chapter, we use *PI* to refer to either a PI or an in-lab mentor who might be a grad student, postdoc, or a staff scientist.

TEN APPLICATION MISTAKES TO AVOID

1. **Applying for an unobtainable position.** Sometimes students are encouraged by a well-intentioned person to apply for every interesting research position—even if the student isn't eligible for the opportunity. This might seem like a good idea, but it can cost time and get your hopes for up a position you're ineligible for. Experienced mentors choose eligibility requirements such as prerequisite coursework, a set number of research hours per week, or a specific academic level because they know what a student needs to be successful in their research program on a particular project. If the position is part of a summer undergrad research program (SURP), requirements might include being a first-generation college student or that an applicant hasn't previously participated in a summer-intensive research program or that a student identifies with a community that is underrepresented or underserved in science, technology, engineering, math, and medicine (STEMM). **Read eligibility requirements carefully because they won't be waived, not even if you apply with a stellar essay and outstanding recommendation letters.** However, *note that if a particular qualification is preferred or even strongly preferred (but not required), it might be worth completing an application*, even you don't have a certain prerequisite.
2. **Sending a generic, inauthentic, or problematic email.** The email you send can be the most powerful, effective method used to secure an interview, or it can immediately ruin your chances for a particular research position. Although some mentors might let you know if your email ends up in one of the categories listed, most won't. We've dedicated a section to composing emails that push through the hidden curriculum filters later in this chapter under the heading "First Contact by Email."
3. **Not demonstrating specific knowledge about the research program and a genuine desire to learn more about it.** PIs know that if you're genuinely interested in a research project, you'll have a more meaningful and rewarding research experience. You'll also be more successful and more likely to meet challenges with the determination to solve problems. **Unfortunately, many students who would be enthusiastic researchers send the opposite message when they first contact a PI. Common mistakes include these:**
   » **Sending a generic email.** Anything along the lines of, "I read about your research and thought it was fascinating," or "I read your research advertisement, and I'm intrigued," is generic and rings inauthentic. Using those phrases is also a lost opportunity to get your email noticed because they are overused. Being specific is essential. It's not enough to

say that a research opportunity is fascinating—you need to state *why it's fascinating to you.* **The best way to avoid a generic email is to include a customized Impact Statement in your email.** An Impact Statement is a short statement (one or two sentences) that demonstrates what you've learned about the PI's research program and that you're genuinely interested in being a member of their lab team. Whether you state the specific reason a project or research program is interesting to you or what you hope to gain from working with members of a specific research group, customization is the key. Tips on writing impactful Impact Statements are covered later in this chapter, in the section "Step 2: Make First Contact."

» **Applying for the wrong position.** If you apply to a research group that studies cell signaling during development but state a passion for barn owl ecology, or you apply to a lab that specializes in cancer drug screens but state that your dream is to study marine mammal population dynamics, it's clear that your interests don't match the PI's research program. **A PI doesn't want you to join their research group if you're not interested in their research, so most will pass if you state *significantly* different research interests from theirs.** This mistake is also relevant when applying to SURPs. The research interests you cover in your essay shouldn't be substantially different from the research projects you select elsewhere on the application because this is one reason that some committees reject an applicant.

» **Judging research positions as interchangeable.** Avoid using any statement similar to, "I'd love to participate in your cutting-edge research program because it's exactly what I've wanted to do since I was in high school, but if you don't have an available project, can you refer me to someone who might? I need to get started this semester." Students often use a similar statement in emails thinking that it demonstrates pure enthusiasm and excitement for research. Unfortunately, it often sends the message that an undergrad believes that all research projects are the same or they aren't picky about what they work on as long as they get into *any* research program. To some PIs these statements also send the message that a student might be searching for a recommendation letter, not a research experience, and the student might bail at the end of a semester after a letter is submitted. Again, PIs often choose undergrads who want a specific opportunity because the undergrad feel a connection to the research or value what they can learn from a project or topic. But if a PI doesn't have an available research position, PIs will often, unprompted, forward a student's email to a colleague who works on a similar topic if the student demonstrates self-directed learning and genuine enthusiasm.

» **Asking the PI for an appointment to explain their research program or give a tour of their lab.** On the surface, this seems like a good idea. After all, research programs are complicated, and who better to explain the details than a PI? Therefore, it can be tempting to send an email similar to, "I'm fascinated by what you do, can you tell me more about your research or give me a tour of your lab?" The problem with this approach is that you miss the opportunity to demonstrate both your desire and your ability to learn on your own—two things that most PIs believe are essential. You also risk sending the message that you might be a do-the-minimum kind of student, which isn't a competitive strategy. Obviously, this isn't the message you intend to send, but it's the one that is often received with this approach. You can avoid this misunderstanding by using your customized Impact Statement when you first contact the professor, and then if you want to ask for a tour of the research facilities, go for it.

4. **Overemphasizing certain personal development goals.** Every PI knows how valuable a research experience can be for an undergrad to develop, refine, and demonstrate interpersonal development. However, many PIs hesitate to interview to a student who emphasizes certain interpersonal goals during the application stage. **In particular, when applications are completed through email or an online form, some personal development goals can be misinterpreted.** For example, stating, "I would like to learn some better methods for concentrating when something is boring," sends the message that the student might not actually be interested in the PI's research but that the student hopes to train themself to focus on the project tasks. As you establish mentoring relationships during your research experience, your personal development goals, and how you can achieve them, will likely become part of the conversation. But at the application stage, *especially through email*, most *interpersonal* development goals are best left unstated unless you're specifically asked about them.
5. **Ignoring the application instructions.** Because everything from you and your labmates' safety to the conservation of expensive supplies and (sometimes) irreplaceable samples relies on a researcher's ability to follow instructions, some PIs use an application to identify students who are already pros at following instructions. Additionally, some processes require that supplemental materials (a personal statement, a curriculum vitae [CV] or resume, a transcript) be sent or uploaded before an application is considered completed. In these cases, if a student doesn't follow through with providing the requested materials, they won't be eligible for a position because they didn't complete the application. (Many SURPs function this way.) So, if an application requires a copy of your unofficial transcript, and you don't know how to download it, ask someone to help. Or if you're

asked to send a resume, don't email the PI to state that you didn't have one but are submitting the application anyway; that's a waste of your and the PI's time. Instead, take the time and effort to compose and submit a well-organized resume to fulfill that requirement. **Always take the time to read each advertisement, application, and email from a potential mentor thoroughly.** Give complete attention to the instructions to ensure that an avoidable mistake doesn't prevent you from getting an interview or a position.

6. **Delaying a response when asked for more information.** One of the ways to demonstrate enthusiasm for a specific research position is to respond to requests for more information quickly. Therefore, once started, it's a mistake to postpone completing the application process for too long. **If asked for additional information** (such as a transcript or resume) **or to take additional action** (such as to fill out an online application), **follow up within one day** (or two, maximum), otherwise you might put yourself out of consideration. Even if you respond after a week with an explanation for the delay such as, "Sorry, I had exams/was sick/had my parents in town/etc.," the reason won't matter if in the meantime the PI scheduled interviews with other students and filled the position or if the application deadline has passed. Therefore, respond quickly so you don't risk losing the opportunity to another candidate by default and to demonstrate that you're enthusiastic about the position.
7. **Presuming you'll earn a salary or class credit for conducting research.** Don't presume that the experience you're applying for will have the same benefits as the one a friend is in. Although some research positions are paid, unless one is advertised as such, you should presume it's a volunteer opportunity or offered for class credit only. If there is no indication of potential compensation on the PI's website or in an advertisement, you'll need to clarify this matter before accepting the position. But don't misunderstand us. **There is nothing wrong with making a paid position an essential criterion for choosing a research position.** And many undergrads who have jobs while in college (as we both did) need to prioritize research positions that include a paycheck.
8. **Appearing to struggle with time management or overcommitment.** A student once told me (DGO) that their biggest challenge in research would be to "not prioritize my research project if I get behind on studying." The student intended to demonstrate a dedication to research, and at first this might seem like a PI's dream. However, the message they sent was that they anticipated having trouble managing their time. **Most PIs will hesitate to offer a position to a student who appears to be overcommitted or is unsure how they will handle their academic and life balance.** There is too much risk involved, and if it doesn't work out, it's not good for

the student, and it's not good for the mentor. (If you haven't considered your schedule yet, return to the chapter 4 section "Step 1: Schedule and Prioritize Your Time" as soon as possible so you can answer scheduling questions with confidence and honesty at an interview.) It's not easy to balance academics, and research, a job or an internship, and family and social obligations, but many mentors will hesitate to offer a research project to a student who is unsure of how they will manage their time or indicates that they are currently struggling to do so.

9. **Intentionally misrepresenting anything during the application process.** Before you include something on an application or CV, ask yourself this: "Is it truthful? Is it accurate?" If it's not, don't include it. A PI must be able to trust each person in their research group to work with integrity when recording results and data, when conducting research, and when working with labmates. But if someone intentionally misrepresents themself on an application, will they be forthcoming when they make a mistake or break something, or will they try to hide the evidence? Will they be honest when they don't understand a safety procedure or how to use a piece of equipment, or will they instead decide to move forward on their own, possibly putting labmates at risk or invalidating a procedure? If a PI doubts your character or credibility, they probably won't call you on it, but they definitely won't offer you a position with the research group, either. **Some examples of misrepresentations** in this category that we or colleagues have observed are these:
   » **Easily verifiable information.** This includes a transcript, major, and academic year. If certifications, vaccinations, specific training, or a financial aid award is required, proof will likely be requested prior to the start date and sometimes at the interview. If, for example, a work-study award is required for the research appointment, and you state that you have one at an interview even though you don't, the PI won't sign off on the hiring paperwork because later you claim to be mistaken.
   » **Work history, volunteer experience, or skills.** Even if a PI doesn't check on employment dates (most won't unless you're applying for a paid position that requires specific, previous employment), confirm volunteer activities, or contact references prior to an interview (most PIs don't ask for references), everything listed on your CV is fair game for discussion during the interview. **Usually, it's immediately obvious when items are inflated if an interviewee is unable to answer detailed questions about the experiences listed on their CV.** One colleague declined to hire a potential undergrad technician after the student claimed to be responsible for ordering lab supplies in their current position but it was revealed that the student had only placed a single order. Another colleague in a similar situation declined to hire

a candidate for a professional research staff (PRS) position after the candidate misrepresented only observing a labmate performing a technique as "copious experience" conducting the technique themself. In both cases, our colleagues shared that the experience the candidates claimed to have was advantageous but not the only reason the applicants were offered interviews. However, because each candidate substantially exaggerated their experience level, and because other candidates applying for the positions didn't misrepresent anything, our colleagues made offers to those other candidates. For you, the lesson is this: if something isn't true, leave it off your CV, and when something is true, don't exaggerate your experience level or other details. All other issues aside, you don't want to start a research experience wherein you've misrepresented your expertise to the PI because then they won't plan to train you in something when you've assured them that you already know how to do it.

» **Proposed time commitment.** This includes a good faith estimate of both the hours per week and the total number of semesters you plan to commit to the research experience. It will be awkward in an interview if you're offered the position and then say, "Oh, by the way, I can't be here in the summer even though the advertisement stated it's required," or "I can't commit to the number of hours I stated in the email." At best, you'll annoy the interviewer if those time commitments are required to be successful in the position. At worst, they might immediately rescind their offer (especially if that time commitment is essential for your success), and they probably won't consider you for a future position with a lower commitment level or refer to you a colleague. If between applying and interviewing your anticipated time commitment for research has changed, send an email to the interviewer right away. They might or might not care, but they should be informed. And take note: if you're thinking that the work-around for this is to apply to a three-semester position but plan to leave after one, know that this approach could lead to some negative consequences at that time, as discussed in chapter 3.

» **Intended career path.** Although this isn't always the case, typically PIs don't decline to offer a research position to a student because of their long-term career goals—after all, those could change over the course of a college career. However, **undergrads who declare one path but are obviously preparing for another could lose out on a research position at the application stage.** The most common example is a student who declares a pregrad or undecided path when they are obviously premed. If you've been told that some PIs won't mentor premed students, there is no need to be concerned because if you're premed, you don't want to end up in one of those research experiences anyway. We covered this topic in depth in chapter 3.

» **Having read a paper.** Some undergrads feel pressured to claim that they have read one of the PI's published journal articles even if they haven't. Don't give in to this pressure. We've already covered why it won't help you in chapter 4.

10. **Not scheduling time to get the applications done.** For many undergrads, around the time they are ready to send out applications, the excitement connected to conducting research starts to wane. It's not that they've changed their mind but that the process starts to dampen their enthusiasm. If this happens to you, dig deep to stay self-disciplined to finish it up—don't put off the application process. The strategy in this chapter will help you get through this as quickly as possible and submit high-quality applications.

APPLICATION PROCEDURE

There is no standard procedure to apply for a research position, and the process can even vary within the same research group. An application might be as informal as "email [a specific member of research group] a transcript, resume, and short statement that explains why you want to conduct research with us," or it might include an online application with several short-answer questions or an essay.

In some research groups, the PI chooses all undergrad researchers, while in others the in-lab mentor who will work most with a particular student drives the selection process. Each PI (or group member interested in mentoring an undergrad researcher) determines the selection criteria, application process, and interview procedure that works for them. With SURPs the selection committees or directors create an application and procedure that best serves a particular program. This nonuniversal application procedure can feel frustrating to you as an undergrad, but if you want a research experience, you'll need to push through those feelings and get the work done.

Regardless of the application procedure—whether done through email or another method—it's important to remember that you could disqualify yourself for an opportunity by not following all the instructions.

Continue to use the hours you scheduled for a research experience in chapter 4 to complete the application process. This is important for two reasons: (1) it will help you prioritize finishing your applications; and (2) you'll continue to test whether you can commit to your proposed research hours in a research position. All PIs prefer undergrads who can manage a research commitment without compromising their academic and life balance. **Stat-**

ing that you can manage a research commitment makes less of an impact than stating *why you know you can*. When you explain how you've determined your available hours at an interview (from scheduling them to giving them a "road test" as we recommend), you'll set yourself apart from the students who can't.

STEP 1: CRAFT YOUR CV AND OBTAIN YOUR TRANSCRIPT

As an undergrad applying to your first research position, you may find that your accomplishments might be more appropriate for a resume than a CV. However, as you take advantage of the professional development opportunities available through an in-depth research experience, you'll probably find that CV will become the more accurate label. At the start, however, **it doesn't matter what you call it—CV or resume—unless you apply to a research position from an advertisement that specifically requests one or the other.** In that case, label your document with whatever is instructed. In the interest of brevity, we'll use the term *CV* to represent both document types for the remainder of this book.

Although there is nothing exciting about crafting your CV, many PIs use one as part of their application procedure. Therefore, you need one.

Fortunately, it's relatively easy to create a high-quality CV from a template that is provided with most word processing programs or can be downloaded for free from the internet. As a student, you might have access to a free version of Microsoft Word through your institution or already be using Google Docs, so these might be good options. To some potential mentors (and most employers), a poorly constructed CV sends the message that you were either too busy (not a good sign) or too lazy (even worse) or unable (yikes!) to produce a high-quality one. **As tempting as it is, don't cut corners here.** Although a less-than-perfect CV won't prevent you from getting an interview, **a well-crafted CV will give you the advantage over students who submit a subpar one.** One story that underscores this point this was shared by a colleague who had an open staff scientist position in their group. A candidate submitted a CV that was clearly unfinished as it had notes in the margins that included changes suggested by someone who had reviewed the document for them. Whether the candidate accidentally sent the wrong version of their CV or presumed that the person helping them craft it would make changes and not comments didn't matter to our colleague. With so many people applying to the position, our colleague declined to consider this candidate who didn't appear to be detail oriented at the application stage. In another research group, the PI might inform the candidate of the unfinished CV and ask them to resubmit it after they finish editing it. For many undergrad research opportunities this might be more likely to happen, but you shouldn't count on it.

A CV is so much more than a compilation of your accomplishments and activities. It represents you. It showcases your achievements and demonstrates that you believe details matter.

THINGS TO REMEMBER WHEN CRAFTING YOUR CV

Before you start, keep in mind that a CV distills a vast amount of effort into a few lines of text and almost always makes its creator feel inadequate. So, don't worry about what your friends put on theirs or feel discouraged about what you have (or don't have) to put on yours. **As you create your CV, remember that it takes time, effort, and experiences to build a CV.** Few students have a distinctive one at the start of their college career. A research experience will present numerous opportunities for professional development and will help yours grow.

Also, know that experienced researchers understand. Most PIs won't make their decision to interview based on your activities, accomplishments, or skills (unless a research project has specific requirements). And they won't judge you on a personal level for the number of activities and accomplishments on your CV. After all, they have been there and know that it's just a fact that opportunities build CVs—not wishful thinking.

TWELVE TIPS FOR CRAFTING A CV

Your goal is to create a CV that includes the information a potential mentor will likely want and to present it in a polished style. Follow these tips on how to put together a CV that will highlight academic accomplishments, skill sets, and experiences. Most of these tips also apply when you're crafting a CV for a scholarship, fellowship, job, or other application.

1. **Find a template to customize.** Most word processing programs have a CV template you can quickly customize with your information and accomplishments. Alternatively, do a web search for free templates. Make one CV to use for most (ideally all) research applications and use it as a foundation to build on for the rest of your college career.
2. **Add your name in the file name.** It's possible that the person who reviews your CV will download it to their computer at some point. A generic label of "CV" isn't helpful when they want to find *yours* quickly. When you save your CV, always put your name in the file name. A good label to use is "[your name] CV."

3. **Including an "Objective" section is a matter of preference, but it won't count against you to leave it off.** If you want to use the same CV for all applications (which we recommend), you could inadvertently give the wrong impression by using a too-specific objective. If you state that you're "seeking a research position in a neurobiology lab," and then you apply for a position in a lab that doesn't do neurobiology research, then the PI might decline to consider your application rather than risk your being disappointed with the group's research focus. Or if you state, "seeking a research position in Dr. X's lab," but then you accidently send your CV to Dr. Y's group, you probably won't get an interview. Also, it might be difficult to believe, but you'll need a CV more than you can imagine in college (we hope that you'll apply to research-related scholarships, fellowships, and travel awards), and it's an added burden to update the objective every time.
4. **Aim for one page in length.** One page is long enough for almost every undergrad, even if you already have a lot of life and work experiences before starting college. But if you decide to make your CV two pages, include your name in the header of the second page and add a page number. Unless you're a student with an extensive work and life history, don't create a three-page CV. It's just too much for an undergrad research experience, and it's doubtful that someone will read past page two.
5. **Summarize academic accomplishments.** If you're early in your academic career, you might not have much college "stuff" to include on a CV. Consider adding a section of "Academic Accomplishments" near the top with items such as the dean's list, any fellowships or scholarships you've received, a list of instructional lab or field classes you've taken, or challenging science classes you've completed. This list might be a summary of your transcript, but that's okay provided that it's a short section.
6. **Include nonresearch activities and accomplishments.** No PI will discard your application because you list accomplishments unrelated to research or because you include a few awards from high school. On the debate team? Won a scholarship? Placed at a science fair? Participated in a sport? Add a few items. Just make sure to place them under relevant headers. One student's CV that I (DGO) reviewed listed "traveled to France" under "Skills." This didn't prevent them from getting an interview in my lab, but for someone else it might have been a bigger issue.
7. **Include all work experience.** Some PIs place a high value on volunteer service or nonresearch work experience done in high school or college. One colleague, for example, is more likely to offer a research position to any undergrad who has worked (or currently holds a job) in any realm of the service industry. So, include your work and volunteer history on your CV—whether you've volunteered at a hospital or an animal shelter or worked in retail or fast food, or elsewhere. Work or volunteer experience

won't count against you even if they aren't research and might even help you secure an interview.

8. **Proofread.** Use both the spellchecker and the grammar checker to help catch the most obvious errors, but don't rely on these tools to catch *all* errors, and be judicious about accepting their suggestions; these algorithms often fail to account for the nuances of the English language and sometimes can actually lead you astray. You'll still want to ask someone else to review your CV before sending it off to a potential mentor (covered next).
9. **Use professional feedback to polish your CV.** When you're satisfied with your CV, it's time to ask for a second opinion from a professional at your campus Career Resource Center, Writing and Study Center, or Office of Undergraduate Research, if those are available. You can also ask an academic advisor. Although a friend might volunteer to check over your CV, they also might be too shy to point out errors and don't have the experience or knowledge to offer professional feedback. For most people, unless they enroll in another academic program after their undergrad years, accesses to these free career resources ends after graduation. Take full advantage of these campus resources while you're still an undergrad. Remember, you're creating a CV to use as a base for future CVs.
10. **Only email your CV as a PDF (portable document format, ending with the file extension .pdf), not as a text file (extension .docx) or a Google Doc.** You'll keep the original text version around to update and revise as your CV grows, but before sending out your CV as part of an application, convert it into a PDF. Files in PDF form will generally appear the same on any operating system, whereas Microsoft Word or other software text files (.doc or .docx) or Google Docs may appear differently. You don't want your CV to appear poorly formatted because a PI doesn't have the latest version of Word. When you send your CV as a PDF, it eliminates that possibility. Typically, you can make a PDF of a Word document by selecting the Save As option for .pdf in the Print dialog box. If not, check in with the computer specialists on your campus or do an internet search to find instructions—this is something that you need to know how to do.
11. **Email a copy of your CV to yourself.** Before you send your CV to someone you want to impress, email it to yourself and then open the attachment to double-check that the PDF file is formatted as it should be. If it needs adjusting, it won't take long to do it. Don't waste the effort you spent crafting a high-quality CV only to have formatting hiccups (such as an extra page with only one character on it) appear when it arrives in a PI's email inbox.
12. **Save both versions in the cloud.** Save both the Word document and the PDF versions of your CV in the cloud. You probably already have an app you use for similar purposes, but if not, choose one of the many free

options. Also, label the storage file "CVs" and resist the urge to store anything else in this file. This way, you'll maintain an archive to select from as you customize future CVs. It takes less than five minutes to upload these documents to the cloud, and if you take the time to do it now, your future self will be grateful more times than you can imagine.

DOWNLOAD AN UNOFFICIAL TRANSCRIPT

Download, or do a screen capture of, your transcript (unofficial is fine). Even if you're a first-semester student, you'll have a transcript—it just won't list much. **Convert your downloaded file or screen capture to a PDF and put your name in the file name ("[YourName]Transcript.pdf"). Ensure that the entire transcript is there**—sometimes information is cut off during the download or screen capture, or the personal information (your name, major, academic year) is inadvertently left off. If a PI or SURP uses a transcript as part of the application process, they might want the full transcript to evaluate the complete academic picture of the applicant. Even though it may seem like a small detail, for some research applications, sending an incomplete transcript means submitting an incomplete application.

STEP 2: MAKE FIRST CONTACT

This step is most relevant to students who are applying for research positions on their campus. If you're applying to a SURP, some of this information is applicable to your situation, but you can also opt to skip ahead to the subsequent section "Step 3: Complete Online Applications."

Now that you've crafted a CV, downloaded your transcript, and identified ten to fifteen interesting projects or research labs, it's time for what we're calling *First Contact*. **First Contact is the first time you contact a PI to ask for a research position.** First Contact can be done in person, through email, or by submitting an online application to a database. Why does First Contact matter? Every PI, grad student, and professional researcher is busy, and many don't have the luxury of time to interview all students who inquire about a research opportunity. Therefore, **in the few seconds to short minute that a potential mentor considers your request for a research position, they evaluate three key questions:**

1. Where does this student rank on the "professional scale"? (Remember, this is code for the hidden curriculum on expected ways to interact with others in email or in person and is generally decided by the first impression and overall impression.)

2. Does this student demonstrate a genuine interest in what I do? (Methods for effectively demonstrating this are also part of the hidden curriculum. A well-written Impact Statement generally covers you here.)
3. Does this student have the prerequisites I require? (These could be a compatible schedule, specific previously completed coursework, or other requirements.)

In other words, as the adage goes, *you never get a second chance to make a first impression*, and **First Contact is the quintessential first impression. This is your first real test.** Do well and you'll practically guarantee an interview if the potential mentor has an open research position.

The best way to contact a PI is a matter of opinion, related to your comfort level and the individuals involved. If you decide to make First Contact by email, and you customize the templates we've provided, you'll probably be able to send ten emails in the time it takes to drink a venti latte. If you decide to make First Contact in person, it will take longer, and you'll want to first prepare by reading the tips in both this chapter and chapter 6.

FIRST CONTACT BY EMAIL

It's true that professors receive a lot of email—sometimes more than is humanly possible to read. However, it's an oversimplification to say that high email volume is the *only* reason a PI doesn't respond to a student's inquiry about a research position. Most professors at least skim the emails they receive from students and try to prioritize answering inquiries from undergrads who are enrolled in the classes they teach.

Although there can be several reasons an inquiry from an undergrad searching for a research position goes unanswered, in this section we demystify reasons connected to the hidden curriculum. And some potential mentors (including PIs, grad students, postdocs, and staff scientists) are aware that an undergrad might not know the expected phrases to use when making First Contact through email, but you won't know the pet peeves of anyone you contact, so you'll want to opt for the more formal approach in most cases (the approach that navigates much of the hidden curriculum connected to emails).

If you've struggled to find a research position and have already sent twenty-plus unanswered emails, the problem is probably that your approach doesn't align with the hidden curriculum. The good news is that if this is your situation, you can do something about it moving forward. But even if you're only getting started on connecting with potential mentors, to make the best First Contact impression through email, follow the advice in the next sections.

Twelve Tips for Writing Emails to Navigate the Hidden Curriculum

Each email you send represents you. So, **always email with distinction because the details matter to some recipients**, and through email you can establish yourself as someone who also cares about the details. Plus, it's highly unlikely that you'll offend someone by sending a more formal email.

The tips here (excluding the research-specific items) apply to all emails you send to professors or administrators—whether you are applying to a research position, need clarification on lecture material, or want to schedule an appointment with someone. The tips in this section are intended to increase your chances of receiving a reply email and in some cases can help set the stage for a possible recommendation letter later.

1. **Use your official college or university email address.** Email filters can send nonuniversity emails directly into the bulk, spam, or trash folder. Even if it's inconvenient, use your college email address to make sure your email arrives at its intended destination.
2. **Include a relevant subject line.** Short (fewer than sixty characters) is particularly helpful if the PI reads email on their phone. If you apply for a research position from an advertisement, put the name of the research project in the subject line. If you apply after attending a seminar, put "Caught your seminar. Undergrad research position?" in the subject line. If you apply after attending a poster session, put "I've read your research interests. Undergrad research position?" in the subject line. These highly specific subject lines can set you apart from generic ones. But if you choose a general subject line, make it simple, such as "Seeking undergrad research opportunity."
3. **Use a proper salutation, and spell the person's name correctly.** The salutation you use in an email matters. It sets the tone for the email, and it demonstrates respect for the person you're emailing. For better or worse, it also influences your reader. And you don't want the person reading your email to feel annoyed or offended right from the start—especially because your plan is to ask them for a research position. When emailing about a research position, avoid using an informal greeting such as "Hi," "Hello," or "Hey," or leaving it off entirely. Address the person by their name and title as "Dr. X" or "Professor X." If you're emailing a grad student or staff scientist who doesn't have a PhD degree, most won't be offended if you opt for "Dr." However, numerous professors, and others, will take offense if you use "Miss," "Mrs.," Ms.," or "Mr." Plus, when you choose those titles, you risk misgendering someone. If you're unable to determine someone's professional title, you could also use "Dear [First Name Last Name]:" Even if you and the PI email back and forth several times, continue to use a

professional title in all email communication. (After you join a research group, your mentor will probably ask you to drop their title when addressing them.) Also, ensure that you spell the person's name correctly because it's about basic respect. This is a hot-button issue for many people, and we know of a few colleagues who no longer respond to emails in which their name is misspelled.

4. **Customize the email—don't send a generic one.** Avoid using bcc and cc when contacting multiple PIs about research positions. Mass emailing as many people as possible sends the message, "I'm searching for *any* research experience," and it can be a lot of effort for little return. For some students, this approach might result in an interview, but for many this "spam-a-lot" approach will be unsuccessful. Being on the receiving end of spam email is the same for a PI as it is for you. No one likes it. Plus, the mass email approach is often ineffective because it doesn't allow you to customize an email for each research position. The result is a vague, generic email that appears to be sent to a bunch of random people (which, in a way, it is). It's impossible to convince a PI that you're genuinely interested in *their* research program if the email is also addressed to twenty other people, or it could have been sent to everyone as is. Therefore, avoid using bcc and cc or sending generic emails.
5. **Get to the point.** As you know, PIs receive a lot of email, so they appreciate brevity. Start each email by declaring your interest in a research position and why you're specifically interested in the PI's research program. Keep the email body short. For some potential mentors, the longer your email, the less of it that gets read. This may already be familiar to you: the longer a class syllabus, the less of it you probably read. And think about the time you've spent reading this book. In places, you've only read the words in boldface. You're busy. You prioritize. PIs are busy. They prioritize.
6. **Avoid filler words.** This is an extension of the previous tip. It's common for students who apply for a research position through email to state that they are they a hard worker, a fast learner, and filled with enthusiasm for science. Unfortunately, in part because these filler words are overused, such promises don't make much of a distinctive impact on experienced research mentors. Filler words don't do much other than lengthen your email, so skip them. By incorporating an appropriate Impact Statement customized for the research group, you've already shown more self-directed learning and enthusiasm than filler words could ever do.
7. **Include a separate document describing your previous research, if applicable.** Even if you've participated in research at the college level, a lengthy explanation of it is unnecessary in the email body. At most, add one or two sentences to your email such as, "I have previous research experience working with *X*. I've attached a PDF describing that experience."

In this PDF, which should be somewhere between 150 to 500 words, cover (1) the number of weeks and average hours per week you participated in a research experience; (2) a short explanation of the project and how it related to the research group's big picture or a background summary of why the research you did is important; (3) what you gained from the research experience or what your favorite part of it was; and (4) some details about your specific responsibilities. When you save your research statement, put your name in the file name, such as "[YourName].PreviousResearch.pdf."

8. **Avoid including a *long* personal backstory in the email body.** Maybe you have a personal connection to the disease a lab studies or are interested in advancing research on pollution and its impact on your home community. It's understandable to want to share why you identify with a certain research program on a personal level, but to avoid adding significant length to the email, include it in your Impact Statement instead. Then attach a PDF in which you share more details on the subject, just as you would for previous research experience, but label this document as "Inspirational Statement." This introduces your backstory to the person reading your email and gives them the opportunity to read more about it but doesn't make your email too long for others who might not want this information in an email.
9. **Avoid flowery language.** Sometimes, while trying to express enthusiasm for a research position, a student will use flowery language in their email. This approach can also backfire. There are three basic reasons to avoid flowery language. **First, it's distracting.** Flowery language takes away from the message you want to send, which is that you want to learn what the PI has to teach, and you have the enthusiasm and prerequisites to make a contribution to their research program. **Second, it's highly annoying** (to some). Some PIs find flowery language so irritating that they'll quit reading an email as soon as it appears. **And third, you don't need it.** You don't need flowery language to prove that you're smart—your admission to college already does that. And you don't need it to prove genuine enthusiasm or the ability to learn on your own—your Impact Statement will do that for you.

   Here's how to recognize flowery language.

   After you write a draft email, set it aside for an hour or until the next day. After you return to it, read it again and consider how conversational the words seem. Then **ask yourself this: "If I asked a professor in person or in an email about lecture material, would I use similar words and phrases?" If not, your email might be too flowery.** It can be difficult to let go of a beautifully crafted statement that you worked hard to write, but if it has flowery language, set it aside to use elsewhere because using it isn't necessary to get a research position.

And remember, a straightforward Impact Statement will be the most effective tool for driving home the basic message of "I want to learn what you have to teach, and I want to contribute to your research program."

10. **Check the spelling before sending.** You've been told it before because it matters so much to some people. Inquiring about a "reserach opprotunity" doesn't make a positive impression. It's also enough for some PIs to decide that a student might not be detail oriented or doesn't think that an impression made through email is important. (Remember to spellcheck the subject line, too.) Also, when writing your emails, use full words, not just letters or numbers: *two* not 2, *be* not *b*; use correct capitalization: *I* not *i*; and skip the emojis. To you, these suggestions might seem overly critical (and some PIs would agree), but your goal is to write emails that appeal to the broadest range of possible mentors.
11. **Prevent unwanted font changes.** Every word in your email doesn't have to be unique—the salutation and your Impact Statement typically constitute enough customization. But you'll lose credibility if you send out an email that contains multiple fonts, because it will appear as if you sent the same email, with only a slight modification, to several people. That puts your email in the bcc category (as explained in tip 4). Even if your email program doesn't show multiple fonts, the PI's might reflect variations. *If you cut and paste any text into your email program, cut and paste 100 percent of it to avoid a font change in the middle of an email.* Alternatively, compose your email in a word processing or text program and cut and paste it all at once or select all text in the email and adjust the text and font size.
12. **Include an email signature.** An email signature is used to relay important information, which helps keep the body of your emails short. Few students create an email signature, so it's an additional way to be distinctive. In your email program, go to the preferences to find the signature panel and create a default signature to use in all email correspondence. Although we recommend including some information, such as your name, in all correspondence, there is some optional information. For example, in this signature, you might choose to include your pronouns if you're comfortable sharing them. Other optional information you might wish to include are the phonetic and audio pronunciations of your name. If you're not sure how to create a name pronunciation key, do an internet search with "how to phonetically pronounce [your name] in [language]." You might also want to consider recording your name in an online database and adding a link to that audio pronunciation in your signature.

    The information to include in your email signature is as follows:

    » Full name
    » Your pronouns (optional)

- Name pronunciation keys or link to an audio recording of your name (optional)
- Major and any minors
- Academic year in college or your expected graduation date
- College or university email address

The Email Template

Each sentence of your email should highlight your genuine interest in, or knowledge about, a research topic or how you'll bring value to a specific research program. Aim for five to seven sentences (eight maximum). When using the guide that follows to write your emails, be flexible if the exact sentence sequence doesn't work for you. For example, you might want to combine the opening line and the next line or ask for a position in the middle of the email. Focus on being specific, direct, and authentic rather than forcing your thoughts into a template.

SUBJECT LINE

This is the first opportunity to show that you did some self-directed learning or put forth extra effort to find a project that genuinely interests you. Expressing these ideas in the subject line also demonstrates enthusiasm.

FIRST SENTENCE

Get right to the point. You're interested in joining their lab or research project, and you've learned something about the research program. (If your financial award packet includes a work-study award, this is a good place to mention it.) You don't need to introduce yourself because you've created an email signature with that information. And you don't need to include information about the classes you've taken unless they are a requirement for the position. You'll attach your CV and transcript, which will detail all your classes and activities, so you don't need that information in the email body. Many mentors only read an email inquiry for a few seconds and then make a decision whether to ask the student for more information, schedule an interview, or eliminate the student from consideration. Make those seconds count!

SECOND SENTENCE

Reference how you learned about the research opportunity or became interested in the research topic. Be specific or it's ineffective. "I ___." Fill in the blank with the appropriate information such as, "attended a seminar," "read a poster," "spoke with [name of person] at a research symposium," "read your research interests," or "found your advertisement for an undergrad researcher."

THIRD SENTENCE

Include a customized Impact Statement. An Impact Statement is a short statement (one or two sentences) that demonstrates what you've learned about the PI's research program and that you're genuinely interested in being a member of the research team. Whether you state the specific reason a project or research program is interesting to you or what you hope to gain from working on a specific project, customization is the key.

Use the three sentences you highlighted about each research opportunity during the search phase to write your Impact Statement, but put most of it in your own words—don't simply copy the text verbatim. Alternatively, if you found a research opportunity through a seminar, research symposium, or other method, use the inspirational statements you wrote down during a presentation as a guide. **There are essentially two types of Impact Statements:**

Impact Statement 1
What you hope to learn from working with a specific research group or working on a specific project.

Impact Statement 2
Why you're specifically interested in the science.

FOURTH SENTENCE

(Optional) Mention your previous research experience, if any. One or two sentences is all you need. Remember, you'll attach a short description of your previous experience in a PDF, so you don't need to spend much space in the email covering it. When someone is interested in learning more about it, they will read the attachment.

FIFTH SENTENCE

Include your anticipated time commitment. First, include the number of hours per week you can devote to research. Second, include the length of time you'd like to participate in research in semesters. If you're applying for a position in the spring and you know you'd like to continue with research through the summer, mention it. If you're not sure about your summer commitments yet, don't bring up the subject in the initial email unless you know that a summer commitment is required or preferred.

SIXTH SENTENCE

(As needed) List specific qualifications. If you respond to an advertisement for a research position, the term *qualifications* will probably be used instead of *prerequisites*. When applying to such a position, specifically mention that you possess each required qualification (but only do this if it's true). This will show that (1) you read the ad thoroughly, and (2) you indeed have the required qualifications. For example, you might write, "As required for this position, I've completed Chemistry 300 and its accompanying lab."

SEVENTH SENTENCE

(Optional) If desired, you can include times you're available to meet to discuss the research opportunities in the lab. This probably won't make the difference between receiving an interview invitation or not, but it might make it easier for a PI to schedule a meeting with you. However, if you do this, be precise when mentioning days and times that you're available. And if you choose this approach, don't schedule anything else for that time slot and don't offer the same appointment time to more than one person at the same time. If you state that you're available, *maintain* that availability because if a PI has the same opening on their calendar, it will be easy for them to schedule the interview with you.

CLOSING LINES

If you want to include an additional line before your signature, make it similar to, "If you don't have an available research position this semester, would it be possible to join the research group as an observer?" but *only* if you're interested in that option. You can also add, "Thank you for your time" or a similar brief closing, which doesn't count against your sentence limit.

Email Templates for You to Customize

To increase the efficiency of your search, we encourage you to email multiple PIs around the same time. But—and this is key—ensure that you customize each email with an appropriate salutation and Impact Statement. Always be accurate and sincere, and don't misrepresent anything in the email. And remember to attach your CV, transcript, and statement of previous research or inspirational statement to the email before sending! Even if you store these documents in an online, sharable folder in the cloud (which we hope you do), *do not* send a link to that folder. Most mentors won't click on an external link to download a document, so include your documents as attachments in your email.

Use the templates presented next to send emails that demonstrate your knowledge of the research program, genuine interest in it, and availability for conducting research.

EMAIL TEMPLATE 1

Use this template after reading the PI's research interests and recognizing techniques that you find interesting or have exposure to from a lab course. Incorporate both Impact Statements 1 and 2: *What you hope to learn* from working with a specific research group or working on a specific project *and why you're specifically interested in the science.*

Subject: I've read your research interests. Undergrad project available?

Dear [Dr. X]:

I would like to be considered for a research position in your lab.

When I read your research interests, I noticed that your lab studies some of the processes that I learned about in my ecology lab. I enjoy collecting and analyzing field data and would like more exposure to this in a research setting. This semester, I have 8 to 10 hours per week for a research experience, and I would like to be involved in research for 2 semesters.

Thank you for your time.

Sincerely,
[First Name]
[First Name Last Name]
Name pronunciation keys (optional)
Pronouns: (optional)
Major: [your major], Minor: [your minor]
Class of 20XX
your.email.address@college.edu
(include relevant attachments)

* * *

EMAIL TEMPLATE 2

Use this template after reading a PI's research interests and incorporating Impact Statement 2: *Why you're specifically interested in the science.* Extra tip: Don't copy directly from the PI's research statements. Put the Impact Statement in your own words to maintain authenticity and sincerity.

Subject: Solar-based fuels rock! Undergrad research position?

Dear [Dr. X]:

From reading your research interests, I learned that you study biological electron transfer to create solar-based fuel. I would like to work on a project related to this in your lab because I'm interested in using technology to help the environment. I have 12 to 15 hours per week available for a research commitment and plan to continue a research project until I graduate. I also have a work-study award.

Do you have any spaces in your lab starting this or next semester?

Thank you for your consideration.

Sincerely,

[First Name]
[First Name Last Name]
Name pronunciation keys (optional)
Pronouns: (optional)
Major: [your major], Minor: [your minor]
Class of 20XX
your.email.address@college.edu
(include relevant attachments)

* * *

EMAIL TEMPLATE 3

Use this template after reading a PI's research interests and incorporating Impact Statement 1: *What you hope to learn* from working with a specific research group.

Subject: I've read your research interests. Undergrad research position?

Dear [Dr. X]:

I would like to be considered for an undergrad research position in your lab.

From reading your research interests, I learned that your lab studies how different skull shapes in vultures relate to their scavenging behaviors. I noticed that your research uses R and Mathematica, and I have some experience in both. I plan to attend grad school in Integrative Biology, and so I want to learn as many research skills and as much about the scientific process as I can. I also hope to contribute enough to earn authorship on a paper. I have 10 hours per week for a research project this semester and will be available for a full-time commitment this summer. Ideally, I'd like to continue with a research project until I graduate.

Thank you for your time.

Sincerely,

[First Name]

[First Name Last Name]
Name pronunciation keys (optional)
Pronouns: (optional)
Major: [your major], Minor: [your minor]
Class of 20XX
your.email.address@college.edu
(include relevant attachments)

* * *

EMAIL TEMPLATE 4

Use this template when answering an advertisement for a research position and incorporating Impact Statement 2: *Why you're specifically interested in the science.* Extra tip: Shorten the Impact Statement if only the lecture course or the lab techniques are relevant. Also, some PIs advertise multiple projects, so always specifically name the project you're interested in joining if that is relevant to your situation. In the subject line of the email, mention the project.

Subject: Molecular Gene Mapping Project

Dear [Dr. X]:

I would like to apply for the undergraduate research assistant position advertised on the Biology Department's website to work on the Molecular Gene Mapping project available in your lab. I'd like to participate in this project because, as I learned in my introductory genetics course, this project will help identify the function of a gene, and I like the idea of discovering something new. Also, I enjoyed doing PCR and gel electrophoresis in my Biology 201 lab, and I know from the ad that both techniques will be used extensively in the project.

I have all the qualifications listed in the advertisement: 10 hours a week for a research experience, self-motivated and reliable, and available for at least a one-year commitment. As instructed in the ad, I've attached a copy of my CV and unofficial transcript.

If this position has already been filled, would it be possible to join your research group as an observer until a similar project becomes available?

Thank you for your consideration.

Sincerely,

[First Name]

[First Name Last Name]

Name pronunciation keys (optional)

Pronouns: (optional)

Major: [your major], Minor: [your minor]

Class of 20XX

your.email.address@college.edu

(include relevant attachments)

* * *

EMAIL TEMPLATE 5

Use this template after attending a seminar, and use Impact Statement 2: *Why you're specifically interested in the science*. Use one or two of the inspiring statements that you wrote down at the seminar. Extra tip: Because going to a seminar demonstrates self-directed learning and making a special effort, you can use the exact phrases you wrote down during the presentation as long as you cite them as your source in the email.

Subject: Attended your seminar! Undergrad research project?

Dear [Dr. X]:

I would like to be considered for an undergrad research position in your lab.

I attended the seminar you gave on [month and day] titled "[Seminar Title]." Although much of it was beyond my academic level, two things that you mentioned caught my interest. First, [fill in topic here]. Second, I really enjoyed [fill in topic here]. If you have an available project at the undergrad level, I would love to start this semester and continue for one year including full-time next summer. This semester, I have 11 to 15 hours per week to dedicate to research.

If you don't have an available project at the undergrad level, would it be possible to join your lab as an observer for the rest of this semester and start a project next semester?

Thank you for your time.

Sincerely,

[First Name]
[First Name Last Name)]
Name pronunciation keys (optional)
Pronouns: (optional)
Major: [your major], Minor: [your minor]
Class of 20XX
your.email.address@college.edu

(include relevant attachments)

* * *

EMAIL TEMPLATE 6

Use this template after attending a poster session at a symposium where you did not have an on-the-spot interview. Use Impact Statement 2: *Why you're specifically interested in the science*. Use the answers you wrote down after discussing the presenter's project to write your Impact Statement. Extra tip: Because going to a symposium demonstrates effort, you don't have to translate the presenter's words into your own if you cite them as the source as demonstrated in this email template.

Subject: Talked with [first name]. Undergrad research project?

Dear [Dr. X]:

I attended the research symposium "[Symposium Name]" on campus yesterday and spoke with [Name of Person] about their project. They told me that they are trying to answer [research question], and that it's important because [reason it's important]. I would like to work on a similar project, as an assistant or as a self-directed researcher, as a member of your research group. I have 9 to 11 hours per week to dedicate this semester and would like to continue with research for at least 2 semesters.

I have 1 semester of research experience as an assistant on a project about [brief but specific descriptor or project topic]. I've attached a short statement covering the details of that research experience, as well as my CV highlighting the skills I gained.

Thank you for your consideration.

Sincerely,
[First Name]
[First Name Last Name]
Name pronunciation keys (optional)
Pronouns: (optional)
Major: [your major], Minor: [your minor]
Class of 20XX
your.email.address@college.edu
(include relevant attachments)

What to Do Next: Your Follow-Through Strategy after Sending Emails

If you've completed the First Contact step through email, don't panic if you don't receive a response right away, and don't spiral into a self-destructive bout of impostor syndrome. Few mentors will respond on the same day, and some will take a week (or more). But what you do next depends on the response (or lack thereof) that you receive.

If you're asked for additional information, or offered an interview, try to follow up within twenty-four hours. A quick response is an easy way to demonstrate enthusiasm for the research position. When they have an available project, a mentor will only schedule an interview with, or ask for additional information from, a student they think would be a strong addition to the research team. Consider this a win. You've made it to the next round.

If you receive a "lab is full" response, reply and thank the PI for answering your email. In this follow-up email, you could ask to join the research group as an observer for the rest of the semester (if it's something that works for you) or inform the PI that you're interested in joining a project next semester (if true) and ask if you should contact them at that time. If you receive no response to this email, then consider it a hard no and focus your efforts on joining the other research groups you contacted through email.

If you receive no response to your initial email after one week, contact the PI again. To avoid any resemblance to email shaming, which won't help you get a research position, forward your original email, or cut and paste it and send as a new email. Your email client might even include a Send Again feature that makes this easy to do. You can politely add a sentence such as, "I was wondering if you had time to read the email I sent last week about joining your research group?" which won't add much length, but you can also leave it off. The PI knows if they read your email and didn't respond. They might simply be trying to get to it, and your gentle reminder will help them prioritize it.

If you receive no response within six days after your follow-up email, either send the email one more time or cross that research group off your list and pursue other opportunities.

If no one responds to your email, or everyone responds that their lab is full, do the following:

- **Review all your Impact Statements.** Make certain that they are relevant, specific, and sincere. If you recycled these statements rather than customizing them, they won't make the impact intended. If you accidentally sent the one claiming eye evolution in jellyfish is your passion to a professor who studies cliff swallow nesting patterns, then you'll know you need to be organized moving forward. (If you did this, it's worth sending the correct Impact Statement to the correct professors in a new email because there is nothing to lose. No need to mention the mix-up, either. Simply start from scratch.)
- **Review the time commitment you proposed.** If you're applying to labs with highly specialized techniques and you only propose a few hours per week or want a short-term experience, that could be the problem. You might need to pursue other types of research programs or increase the amount of time you have for research as long as it doesn't risk your well-being or academics.
- **Review the Ten Application Mistakes to Avoid** presented at the beginning of this chapter to ensure that you're not making mistakes in that list that could be making your search more difficult than it needs to be.
- **Review the "First Contact by Email" section** and pay particular attention to the "Twelve Tips for Polished Emails to Navigate the Hidden Curriculum" list.
- **Review the chapter 3 section "Why Research Positions Are Competitive"** to determine if any apply to your situation.
- **Choose a different approach** for your search from the chapter 4 section "Step 2: Identify Potential Meaningful Research Experiences."
- **Don't become discouraged.** Remember that not all of this process is under your control, and sometimes it takes a couple of rounds to secure an interview.

* * *

FIRST CONTACT IN PERSON

If you would rather connect with a potential mentor in person, there are several strategies to consider. But before approaching someone, write (and memorize) a customized Impact Statement. Use the information and examples in the previous section as a model for how to do this.

Option 1: First Contact during Office Hours

If your professor or teaching assistant (TA) has open office hours (meaning that current and former students are welcome to drop by during posted hours), then simply go to their office or meet virtually during the scheduled time. (You can usually find a syllabus online that lists their open office hours for the current term.) If you're going to their office in person without an appointment, try to show up five minutes early (not much more) or right at the start of office hours because it's more likely that the professor or TA will be fresh and have time to discuss non-class-related topics.

If your professor has closed office hours (only students currently enrolled in a class they are teaching are eligible to attend) and you aren't a current student, then opt for other First Contact methods such as the appointment strategy covered after this section.

Whether you're connecting in person or virtually, use the three tips presented next to make it a successful meeting.

Three tips for attending office hours to ask for a research position:

1. **Arrive prepared.** Memorize your Impact Statement beforehand so you're ready to impress. Also, bring a printed transcript and CV and make sure you have read chapter 6, just in case they decide to conduct an interview on the spot.
2. **Consider your timing.** Although it's impossible to know the best time to attend office hours when you want to discuss an off-topic subject, the worst times are a few days before and after an exam, the first or last week of a semester, or if several students are already waiting when you arrive. So, use the syllabus as a guide to avoid exam time. And although the popularity of online teaching boards and remote office hours have practically eliminated queues of students waiting during office hours, if you arrive and find several people waiting, consider returning at a later date.
3. **Be direct.** If you're currently enrolled in their class, avoid starting the conversation with a fake question about lecture material. Make it clear that your priority is a research position and state your Impact Statement. For example, "Hi Dr. X, I don't have a question about the class material, but I am interested in a research position with your group. I know you study protein trafficking in plant cells, and I'm interested because I'd like to use live cell imaging to understand how proteins move between cellular compartments. Do you have any spaces in your research group starting this or next semester?"

After that, the ball is in your professor's or TA's court. They might conduct an interview, refer you to someone on their research team, make an appointment to discuss the possibility later, or tell you that they aren't currently accepting new undergrads into the research program. In any case, you'll likely either secure an interview or know to cross that lab off your list and continue your search elsewhere.

Option 2: Ask for an Appointment

If you're unable to attend office hours, you can request an appointment from your professor or TA instead. However, whether done in person or through email, clearly **state that the reason for the appointment is to discuss a possible research position.** Be honest about why you want the appointment from the start—there is nothing wrong with wanting to join their research group!

Most people (not just professors) prefer to know in advance what the purpose of a meeting is. If they know that you want to discuss the possibility of joining their research team, it allows them to adequately prepare and to schedule an appropriate amount of time for the meeting (or possibly make arrangements for your potential in-lab mentor to attend). Plus, if someone doesn't have an available project, surprising them with a discussion about it at an appointment won't make one magically available. So, it also saves you time to be direct.

Option 3: Ask after Lecture (but Probably Not Before)

The ability to self-advocate is an essential skill to develop and can require getting out of your comfort zone by asking others for opportunities. Connecting with your professor or TA face-to-face when other students are around and aware of your conversation can be nerve-racking, but it's often the quickest method to determine whether someone has an available research project.

However, you'll want to quickly get to the point because your professor or TA won't have much spare time at the end of a lecture for a discussion. Be bold and direct. **Introduce yourself (unless they already know who you are), state that you're interested in a research position, and give your Impact Statement.** If you're worried that you'll forget what to say in the moment, memorize your Impact Statement ahead of time. It's only a couple of sentences, and no one will penalize you if it seems rehearsed.

After you give your Impact Statement, if your professor or TA doesn't immediately open a discussion about their research program, volunteer to email a copy of your CV and transcript (even if you hand them a paper copy). They might say yes to end the conversation, or they might say yes because they have an available position, and your approach was effective. At this point, their reasons don't matter, but the action you take does.

If they answer *yes* to your emailing the transcript and CV or instruct you to follow up with a member of their research team, then do it that day—ideally within a few hours after class ends, but definitely before you go to bed that night.[1] Following up the same day reinforces your enthusiasm for conducting research with their team. Also, note that as soon your classmates learn that your professor or TA has an open research position, some will contact them and ask to join their research group. So, don't delay sending your email. You should also remind your professor or TA in the email that you spoke with them "after the [topic, such as population biology] lecture this morning and have included a CV and transcript as instructed." If they aren't immediately sure of who you are, this should jog their memory.

Technically, you could attempt to connect with your professor or TA about conducting research before lecture, but there's a higher probability that the conversation will be a nonstarter. Ever notice how some instructors arrive at the classroom with only a minute or two to spare before class begins? Or they wait (sometimes patiently, other times not so much) for the professor from the previous class to get out of the way so they can plug in their computer and get started? They will probably answer as many students' questions as they can while setting up, but a professor won't delay the class start time for a substantial conversation about their research program. So, it might be more effective to ask after lecture.

WHY YOUR PROFESSOR OR TA MIGHT NOT WANT TO DISCUSS A RESEARCH POSITION IMMEDIATELY AFTER LECTURE EVEN THOUGH THEY VALUE UNDERGRAD RESEARCHERS

If you ask an instructor to join their lab after a lecture is over, and they don't immediately say yes but also don't say no, it might be disappointing but it's still a good sign. Even if they are distracted by shoving papers into their bag and quickly reply, "please send me an email," don't interpret that as a negative interaction. Instead, consider that you have cleared the first hurdle and follow up as instructed. Even if they would like to, there are several reasons why your professor or TA might not be able to immediately discuss their research program with you at the end of lecture. **Here are six common reasons:**

1. **They have a packed schedule.** PIs are busy from the moment they wake up in the morning to the moment they go to bed at night. During a typi-

1. Don't connect with your professor or TA before lecture and then email them your CV and transcript during the lecture period. If they notice the email's time stamp, they'll know that you weren't paying attention in class. To some instructors, this would create a negative impression.

cal workday, most PIs schedule more tasks than is humanly possible to accomplish. By the time they finish answering students' questions about course material (reason 2) they might already be behind schedule, or close to it, and need to get moving to their next activity.

2. **They want to focus on answering lecture-related questions.** After class, professors and TAs often prioritize answering questions about lecture material and upcoming exams before heading off to their next appointment. (Some even intentionally schedule office hours directly after lectures, so the material is fresh for everyone.) It's also possible that they are teaching another class immediately after yours in the same classroom (or one elsewhere on campus) and need those few in-between minutes to prepare. Or they might need to clear the room because the class scheduled directly after yours is already funneling in.
3. **It's too important of a conversation to rush**, so they prefer to schedule an appointment where they can focus on the details. For many professors, explaining a research project, outlining their expectations, and answering a potential mentee's questions takes more time than is available in a few minutes after a lecture. Also, if they are a PI, before giving false hope or conducting an interview, they might want to check with members of their research team to ensure that someone is available to mentor a new undergrad.
4. **Someone else in the group interviews potential undergrad researchers.** Not all professors are involved with recruiting or interviewing the potential undergrad researchers to their research group. Therefore, a professor might instruct you to contact another member of their team, such as a grad student, to discuss available research projects. If they do refer you to someone, follow up the same day through email, and *mention that you were instructed to contact them by their PI.* Copy (cc) the PI in the email so they know you followed up right away. (This is an exception to the cc rule.)
5. **They want to determine whether you have certain credentials before conducting an interview.** Many mentors want to review a transcript (at the least) and a CV before scheduling an interview. This could be because they are interested in your grade point average or previous research experience, but it's likely because they want to know if you've taken certain classes or wish to gain a generalized idea of who you are as a student. If they use a CV and transcript to help determine whom to interview, they need more time than is available at the end of a lecture to review the documents. Additionally, some PIs want to assess how genuine a student's interest in joining their research group is before conducting an interview, so they might want to read your Impact Statement in an email.
6. **They want to know that you're serious about research and not just hunting for a recommendation letter.** Generations of students have shared the erroneous notion that persuading a professor to write a recommendation letter is as simple as feigning interest in a professor's research program. And

while it's true that most professors love to talk about their research, it's only when the person asking *is genuinely interested*. PIs who teach large courses, or who have been in academia for a few years, may have explained their research to many glassy-eyed, yawn-stifling students who have no actual interest in joining their research program. (This is usually made obvious when a student asks no follow-up questions about the research program but instead tells the professor that their research is riveting and then immediately asks for a recommendation letter.) If a professor instructs you to make an appointment, follow up with email, or visit during office hours, it simply could be to determine whether you're genuinely interested in their research program. It could also be to learn if you're willing to put forth the modest effort required to send an email or schedule an appointment. Some professors believe that this strategy also helps identify the occasional student who doesn't actually want to conduct research but just wants to list research as a checkmark on their CV before applying to med school.

OPTION 4: DROP BY A PI'S LAB OR OFFICE UNANNOUNCED, WITHOUT AN APPOINTMENT, OUTSIDE OFFICE HOURS

You may have been informed that your best strategy for finding a research position is to be persistent but not be a pest. Although this statement is generally true, it's impossible to know when persistence crosses into pest territory for an individual PI you don't know. Basically, **if you drop by unannounced and the PI is there, you'll meet one of two types of PIs: the one who doesn't mind, or the one who classifies spontaneous visits in the pest category.** That's not to say that the first type will immediately interview you and the second type will show annoyance. It's unlikely you'll learn which category the PI you approach belongs to because, overall, most will be polite regardless of how busy they are in the moment. But you could drop by at the worst possible moment for either type of PI and the one who would normally schedule an interview is in crisis mode because the lab's ultralow freezer is crashing, or a big experiment was just ruined by a faulty incubator thermometer. In either case, if the PI interprets your spontaneous visit as highly inconvenient, you might lose out on a position because they won't have the ability to immediately chat, schedule an appointment, or even take down your email to follow up with you later. And if the PI forgets to follow up with your later, you'll be left thinking that you've been rejected even though that wasn't the case.

We do acknowledge, however, that dropping by a lab or office unannounced does work for some students, some of the time. If you choose this approach, use the same strategy as you would for any in-person interaction: give your Impact Statement and get right to asking for a research position.

But because you won't know how a particular PI feels about spontaneous visits, or even if they will be around when you drop by, our highest recommendations are to go to office hours, email for an appointment, or initially connect with the professor after lecture.

STEP 3: COMPLETE ONLINE APPLICATIONS FOR SURPS AND OTHER OPPORTUNITIES (WHEN APPLICABLE)

Perhaps the most significant advantage of an application requirement for a research position is that it removes some of the guesswork for you. So, when an application is required, instead of thinking, "Ugh, this is an annoying chore," try to think of it as an advantage such as, "I know *exactly* what I need to do to be considered for this opportunity." When an application is required, you know that a specific research project or opportunity is available—especially if the process is connected to joining a SURP. Also, online applications typically provide an overview that covers eligibility requirements, the program's or mentor's expectations, and a substantial description of the available research project or overall program. All these make it easier to determine whether the opportunity is right for you and, if it's not, immediately direct your search to other possibilities.

If you're applying to a scholar bridge program (as described in chapter 4), many of these tips will be relevant to that process. Although some individual mentors use online applications in their screening process, applications are almost always required for SURPs, so the remainder of this section will focus on those programs.

However, **before spending the effort to complete a SURP application, review the expectations for the participants one more time.** When initially making a short list of programs you want to apply to, it's easy for the excitement of going somewhere new to overshadow the reality of doing the actual work. But the way you feel about how you'll spend your time and the type of research you'll be conducting can make the difference between a happy and productive summer or an unfulfilling or miserable one. Most summers we receive inquiries from undergrads who are bored or disappointed with the SURP they joined and who hope that we can help them transfer in to a different lab or program. (We can't.) The reasons for their unhappiness vary, but many mention having selected programs without carefully reading (or ignoring) the expectations or choosing research they weren't genuinely interested in because they thought it wouldn't matter.

Frustrated students have regretfully told us, "I didn't realize that forty hours a week in the lab meant I had to work the whole time," and "I'm basically doing the same techniques or project I did during the semester only it's more boring because it's all day, every day," and "I chose this because it's in [location] and paid a lot, but I don't care about [the actual research] because it's not anything I want to do in the future." In most of these cases, the student would probably have been happier in another program or perhaps spending their summer pursuing other goals instead of conducting undergrad research.

But when you're as confident as you can be about wanting to participate in a particular SURP, start the application process and use the tips covered next to stay on track.

EIGHT TIPS FOR COMPLETING SURP APPLICATIONS

1. **Don't miss the deadline.** Especially for SURPs, few exceptions (if any) are made for late submissions. And if you do miss the deadline, you might be too far in your academic program to be eligible for the same SURP the next year. Therefore, **use the time you scheduled in chapter 4, in the section "Step 1: Schedule and Prioritize Your Time," to work on SURP applications.** These don't need to be marathon work sessions—if you remain focused, you can get a lot done in a few thirty-minute to one-hour blocks of time.
2. **Keep track of important dates.** As soon as you decide to apply, add that program's application deadline to your calendar. Then, add a second date, called a *target date*, that is three weeks *before* that deadline. Consider that target date your final warning to complete the supplementary materials (such as the final round of editing your goal statement), fill out the actual application, or follow up with recommenders who haven't submitted their letters.
3. **Make a list of what each program requires.** Some applications require several types of ancillary documents, such as multiple recommendation letters, a transcript, a CV, a goal statement, or a research essay. Make a list of what you need to complete for each application, so you can check off each item as you finish it. To stay organized during the application process, consider using a task management program.
4. **Send or upload exactly what is requested.** If you're instructed to include a one-page CV, for example, don't include a two-page CV. If you're instructed to send two personal references, don't send five. More is not better. Sending more than what was requested doesn't mean that you went the extra mile or are extraqualified—it means that you didn't follow instructions.
5. **Prioritize writing and editing the application essays.** Writing is hard work. And for most undergrads, writing essays for an application to a research program they *might* be invited to join doesn't rank high on their

to-do list. But it should. Well-crafted, relevant essays are key factors in competitive SURP applications. These essays are a chance to distinguish yourself from other the candidates and to demonstrate that you're genuinely interested in the program. These essays are also your opportunity to address any issue that might, without explanation, make your application less competitive. For example, if prior research experience is *preferred* (but not required) for the program, but you haven't participated in a research project because you've been working to support yourself financially, an essay might be the appropriate place to address this matter. We've included more tips on essay writing in the next section because uploading high-quality ones are that important.

6. **Double-check that you've answered every question, filled in every blank, and uploaded or submitted all requested documentation.** Some online applications won't let you click Submit until all the blanks are filled in and documentation is attached, so there is no possibility of turning in an incomplete application. However, not all applications have such a built-in insurance policy, so if you miss a question, or don't upload supplementary documentation (or later send it in an email as directed), your application might not be considered at all. Although some programs inform candidates when their application is complete or if it's missing a document and the deadline for submissions is approaching, many don't. But it would be a shame if you weren't considered for a research position because of an avoidable mistake in the application process. Therefore, we recommend following up with the SURP coordinator to ensure that your application is complete. For some SURPS, you can submit an inquiry through their website, but for others you'll need find the appropriate person to contact on the application page and send them an email to inquire.
7. **Save the application questions and your answers, if possible.** Prior to submitting an online application, do a screen capture of all the information on the page if you don't already have it in a Word document. This includes the questions, your answers, and any instructions or overview prior to the application portion. This will give you the opportunity to review the information just prior to an interview.
8. **Follow up with your references, if needed.** The people who are serving as your references might submit their recommendation letter by email, through an online portal at the program's website, or through snail mail—the method is determined by each SURP that you apply to. Sometimes, however, your references will need a reminder to get it done. But rather than ask your professor if they have submitted your recommendation letter, first check in with the program's coordinator. If it turns out that a professor hasn't yet submitted the letter, then follow up with them either in person, virtually during their office hours, or through email.

YOUR STRATEGY FOR WRITING IMPACTFUL SURP ESSAYS

As part of the application process, most SURPs require candidates to write at least one essay, but some programs ask for several. It won't take as long to write an essay as it would to compose a term paper, but on average these essays are one or two double-spaced pages in length, so you shouldn't expect to write one in a couple of hours. The potential essay topics cover a broad range of possibilities because each selection committee chooses subjects related to achieving their program's objectives. Essay topics could include describing a previous research experience or current technical skill sets (such as proficiency in programing languages, statistical programs, or field skills) and how those experiences or related skills have prepared you for future success in a specific SURP. Or you might be asked to write a personal statement addressing your STEMM identity and how that perspective has shaped your career path. Or the instructions might include writing an essay about challenges you've navigated in your life. However, at the minimum, most SURPs require an essay explaining why you want to join their specific program—as in what *specific* benefits or *specific* opportunities *this* program offers that will aid you in accomplishing your long-term goals. Regardless of how many essays are part of an application, or what the topics are, our number one piece of advice is to **start the writing process early—as soon as you've decided to apply to a program.** We hope that this start date will be several weeks before the submission deadline (although two months or more is better if multiple essays are required or you're applying to several programs). More time to complete the essay portion of an application provides more time write effective essays. **And because submitting an impactful essay can make the difference between receiving an acceptance letter or a rejection letter, keep the following tips mind while writing yours.**

Bring a Mentor (or Two) on Board Early in the Writing Process

To get started, create an outline of points you plan to address in a particular essay. Do this however it works for you—by writing full sentences or short paragraphs or simply typing out a list of bullet points. Your goals for this part are to start thinking about what you want to share in an essay and then to get something on the screen before connecting with a mentor. You'll need a letter of recommendation (possibly two or three) for SURP applications. These references are also the first people we recommend you ask to provide feedback on your essays. But **to be respectful of a mentor's busy schedule and increase the chance they are available to take on this task, try to ask at least five weeks before the application deadline.**

If you ask too close to the application due date, they might not have time to help, or you might not be able to do a thorough job of incorporating their feedback into an essay. However, if a mentor prefers that you write a first draft instead of the outline we recommend before meeting with you, follow their instructions. Additional options to consider during the essay writing process include consulting with a professional at the career, study skills, or writing center or Office of Undergraduate Research, if available on your campus. Also, even if a SURP submission portal incorporates editing tools or allows you to save your essay and return to it later, we recommend writing in a word processing program. This will make it easy to share documents with mentors and incorporate their feedback.

Use Relevant Examples in All Essays

Using relevant examples probably seems like obvious advice when reading it. But when writing, it's easy to get lost in a story you *want* to share rather than one that illustrates what is *relevant* to share. **Every story you use and every answer you include in an essay should be relevant to the overarching essay theme.** If an essay topic is "Why do you want to join *this* program?" be specific by writing a detailed, expanded Impact Statement. Don't state something similar to "this program is so much better than other programs because this program offers *X*." Instead, frame your answer around what the *X* is and why it's related to achieving your goals. And if you share a story to support a point in an essay, make sure the story is relevant and tie it back to the question. Don't leave the selection committee wondering why you thought a story was relevant—make the direct link obvious.

Also, on their application website, many programs state which topics belong in an impactful essay—such as a discussion of which labs you're interested in joining and why. And some websites include recommendations on what to leave out of an essay—such as details about coursework you've completed. These guidelines are akin to a professor giving you a list of questions that will be on an exam and subjects that won't be included. So, incorporate these suggestions fully when writing your essays.

Resist the Urge to Recycle an Entire Essay

If you select relevant content, you can reuse some material in multiple essays—your professional development goals or career path will likely work in most essays that cover these topics. But avoid using the exact same essay for multiple applications because a detail that is insignificant to you might be enough for a selection committee to reject your application. For example, the name of a specific group you want to work with, their research focus, or even the place where the experience is hosted won't be the same for all programs. And although it's fine to apply to multiple SURPs, especially given how competitive they are, selection committees want to know that your reasons for applying to their program are authentic. If you recycle an essay verbatim, you risk forgetting to change certain details that will give the impression that you think any program will do.

Remove Flowery Language, Use Spellcheck and Grammar Check, and Avoid Filler Words

Earlier in this chapter, we discussed tips for writing polished emails, and many of these tips are applicable to writing essays as well. In particular, review the sections on removing flowery language, using spellcheck and grammar check, and avoiding filler words. Although word processing programs are unreliable for identifying all errors (and can even introduce new errors if not used judiciously), you should still use spellcheck and grammar check before asking a mentor for their editorial advice. This will help ensure that their time is spent making high-value changes, such as flagging an unclear concept, rather than marking misspelled words.

6

Your Interview Strategy

Although it doesn't happen often, sometimes a mentor will invite an undergrad to join the research group after First Contact (covered in chapter 5) and without an interview. However, most undergrads will have an in-person interview or, as is often the case with summer undergrad research programs (SURPs) that require interviews, one conducted virtually.

There is a difference between using an unpolished interview approach (which is understandable at the undergrad level) and an unprofessional one (which rarely results in an offer to join a research group). However, **it's important to note that *sometimes* the phrase "being unprofessional" is actually a cover for discriminatory actions on the part of an interviewer toward a candidate. An example of this unacceptable behavior would be an interviewer declining to offer a research position to an undergrad because the student's identity as a nonbinary person made the interviewer uncomfortable. But in this book, we define a *professional* interview approach as arriving on time and then communicating to the interviewer that you're interested in the research experience by answering questions with a more formal approach than you would use when interacting with a friend.** Many students believe that showing up on time, dressed up, and answering questions with enthusiasm is all they need to ace an interview. But it's often more complicated. For instance, we've interviewed enthusiastic students who were unaware that the positive impression they believed they were making was in reality an indication that they would have difficulties working within a team.

When we asked colleagues to share reasons in the unprofessional interview approach category (as we define it) that led them to decline to offer a research position at the interview stage to any student or professional researcher, answers included interacting with someone who was dismissive, obviously bored, or untruthful; who indicated that the interview was only a formality; or who outright stated that they didn't value the research program for which they were interviewing to join. Although it might be said that these interpretations are open to the interviewer, this section covers actions to take (or avoid) that might accidentally send those messages in an interview.

We also address issues that raise concerns for mentors, who aren't bothered when a student makes a few mistakes during an interview but are unlikely to offer a research experience to someone with an unprofessional approach as previously described. Some of these tips are provided to help dismantle the hidden curriculum, but others address unethical or inappropriate approaches.

TEN INTERVIEW MISTAKES TO AVOID

1. **Not reading this section on the hidden curriculum of undergrad research interviews.** This first section is for our readers who haven't had the benefit of someone to coach them through general expectations at a job interview. Many of the same practices that make someone successful in that scenario apply to an interview for a research experience in science, technology, engineering, math, and medicine (STEMM) as well. For readers who have previous interview experience, these tips might seem like "Interview 101," but sometimes we forget the best-known bits of advice. They include the following mistakes to avoid:
   - **Arriving late.** Also, don't show up thirty minutes beforehand—the ideal window is five to ten minutes early.
   - **Leaving your cell phone on during the interview.** Focus is important. You don't want to be distracted by, or tempted to glance at, a text or social media notification when the interviewer is talking or you're answering a question. Plus, if your phone keeps vibrating, ringing, or pinging, it might distract or annoy the interviewer. So, a few minutes before the interview, enable the correct silencing mode for your cell phone and laptop if you're using it to take notes. We hope the interviewer has done the same with their devices.
   - **Bringing food or drink. Including gum**—no interviewer wants to be distracted by someone working the last flavor crystal out of a stick of gum. Also, your interview might take place in an area where food or drink is prohibited. There is an exception to this: if you manage a medical condition that necessitates drinking liquids, for example, this doesn't apply to your situation. But you may need to inform the inter-

viewer if you're interviewing in a food-prohibited area, although you're under no obligation to disclose a diagnosis.

» **Wearing a lot of fragrance.** Skip the perfume, cologne, body spray, and scented lotion just in case the interviewer has allergies or sensitivities, or if the interview room is small.

» **Interrupting the interviewer.** Yes, you're excited to tell them that they are describing the perfect opportunity for you, but always wait to share until they have finished their statement, or you could quite literally talk yourself out of the research position. Essentially, don't interrupt the interviewer to finish *their* sentences with *your* thoughts.

2. **Oversharing.** One advantage of participating in undergrad research is the numerous opportunities to strengthen interpersonal development and interpersonal skills, as discussed in chapter 1.

Although no reasonable interviewer would expect an undergrad to have polished all interpersonal skills before starting a research experience, avoid sharing information that might indicate a serious need to hone your self-management skills. Even if acquiring those skills is a reason you're interested in conducting research, you don't need to emphasize it at an interview.

For example, I (DGO) interviewed a student who said that their greatest challenge in research would be to overcome their difficulty with working in groups because they didn't value others' opinions. They went on to explain that they hoped to use their research experience to improve their interpersonal skills before needing them in med school. Although it's likely that the student meant to show that they were a self-aware person who understood a personal weakness, the message the student actually sent was that they would be very difficult to work with. Some interviewers wouldn't ask a follow-up question after this statement to determine whether this initial impression was a fair one. Therefore, avoid statements that might raise concerns about self-management, such as these:

» "My mom calls to wake me up in the morning because I overslept for an exam last semester."

» "Honestly, my biggest challenge will be to not pout when things go wrong."

» "If I have to do [research] on my own, it will be hard. I'm used to having a lab partner in my classes, so I didn't have to read all the prelab or do all the steps by myself on the day."

» "My friend said that starting a research project helped them learn to manage their time better, so I thought I'd try it in case it works for me because I'm really disorganized."

» "If research is like the lab classes I've taken, then I hope it will be easier to remain patient during the experiments."

Along these lines, avoid sharing personal observations about the interviewer's appearance or information you've learned about them on social media. Never, for example, should you compliment the interviewer's outfit, clothing accessories, hair color or style, or other aspects of their appearance. Flattery of this sort doesn't belong in an interview. (And for the record, it's beyond inappropriate for an interviewer to compliment your appearance as well. If they do, consider it a "red flag" that they are either unaware of, or uninterested in, maintaining appropriate, professional boundaries in the workplace between themself and undergrads.)

Also, if you choose to investigate your interviewer's personal social media profiles before the meeting (we don't recommend doing this, but we also know that it's difficult to resist), **don't bring up their posts in an attempt to connect with them.** It can feel unnerving and a bit disingenuous when you're interviewing someone, and they launch into a variation of: "I noticed you went to the beach last weekend and ate at Krabie Joz." Or for a stranger to start a conversation about your kids, family, or a hobby or sport you're into.

3. **Not answering a question.** Interviewers want you to answer the question they ask. Although this might seem obvious, many students either try to second-guess what the answer "should" be and give a "clever" response instead of being direct or simply don't answer the question.[1] Evading a question can be a substantial disadvantage if your answer is important to the interviewer.

   For example, I (DGO) might ask: "I have two available projects; one that relies heavily on group work and one that entails more solo work from the start. So, which do you prefer—working by yourself or working in groups?" A student might answer, "I can do both!" But that doesn't answer the question I asked, which was, "Which do you *prefer*?" not "Which can you do?" (I presume that most students can do both.) When I ask this question, it's to help determine which available project would better align with the student's goals: one that requires significant self-reliance and independence from the start, or one that involves working closely with a member of the research team each day. I want to know which they *prefer* so that I can match the undergrad with the appropriate project and spend the rest of the interview discussing the relevant details. Although I've never declined to offer a research position to a student who didn't initially answer the question, others might make a different decision after asking a more essential question.

1. This refers to questions that are appropriate for the position, such as those based on your CV, Impact Statement, transcript, or goals. We're not referring to intrusive personal questions that don't belong in an interview.

4. **Misrepresenting yourself.** You already know why it's important to be honest on an application, but it's equally important to avoid misrepresenting yourself *at* the interview. This is especially important when discussing your professional goals and the interviewer's expectations. In those cases, **telling the interviewer what you think they want the answer to be as opposed to the truth, or how you actually feel, is a poor strategy for finding a research position where your goals are achievable.**

   For example, don't claim that you'll work twenty hours a week for the next four semesters and your dream is to be a geneticist, if you're only planning to be in the lab eight hours each week for one semester, and you want to be a physician's assistant. A principal investigator (PI) could have distinct training tracks for individual career paths. So, misrepresenting yourself to get an opportunity might land you on a project or training path that wouldn't be supportive of your long-term goals. In addition, your mentor can't help you reach your goals if you aren't honest about what those goals are. And if you can't achieve your goals through the available position, you need to know at the interview, not after you've been involved in a research project for two semesters.

5. **Insulting the interviewer, criticizing other people, or making value judgments.** Seems obvious, right? Surprisingly, these problematic approaches are common at undergrad research interviews. Many students haven't had much interview experience before entering college, so they are unaware of the powerful impact negative comments can make, or sometimes unclear on what counts as a negative comment. Part of putting forth a professional image is sharing that you can work as a group member and that you respect others' efforts and opinions. **The most common mistakes that send the opposite message in this category are the following:**

   » **Criticizing anyone.** No matter if you're referring to a lecture professor, teaching assistant (TA), classmate, academic advisor, or former research mentor, it's a risky move to criticize anyone in an interview. **To start, even on a large campus, connections can extend deep—the person you criticize might be a cousin, sibling, spouse, close friend, or valued colleague of the interviewer. And in academia, connections stretch well beyond a department or campus.** A colleague of ours once interviewed a transfer student who complained about the "poor teaching" at their former college without knowing about our colleague's familial connections in the student's former department. On a personal level, criticizing a roommate, sibling, or parent is also a bad idea in an interview. If you share a personal story, make sure it doesn't include how others have disappointed you or been a source of frustration.

- » **Criticizing a previous research experience.** If your first research experience didn't work out, some interviewers will ask why. This can be tricky because you want to be honest (after all, you don't want to end up in the same type of situation that you left), but you don't want to disparage the previous research group or mentor because it won't reflect well on *you*. The best approach is to state something along the lines of, "It wasn't the right experience for me because ___" and fill in the blank with an appropriate reason, such as, "I want to do more benchwork" or "I need a different schedule" or "I didn't connect with the subject matter as much as I thought I would." Keep in mind that the interviewer might contact your previous mentor after the interview to ask about their experience in working with you. And if they do, it's likely that they will ask if the reason you gave for departing the research experience accurately reflects their recollection.
- » **Criticizing a major, career path, or class.** You should avoid disparaging a major, career path, or class at undergrad research interviews. If asked about your choices or an experience related to these topics, always answer with positive (nonjudgmental) reasons for your decisions. For instance, if you're asked why you chose biomedical engineering for your major, give a short statement about why the subject excites you or how the classes are essential for your career path. Avoid language such as, "I chose my major because I couldn't imagine doing something gross like microbiology." You probably won't know the full academic background of the interviewer or all the disciplines that are part of the research group's program. Stay positive to stay on the safe side. Disparaging a career path or major probably won't affect your chances of landing a research position, but it could, so why risk it?

6. **Stating a disregard for learning.** Mentors can *inspire* a student, but ultimately the motivation and discipline to learn comes from within. Thus, most interviewers try to select students who place a high personal value on learning.

   The most common ways students show a disregard for learning in research interviews are by doing the following:

   - » **Devaluing a class.** There are numerous uninteresting aspects to a research experience. What helps most researchers get through them is the knowledge that completing the boring tasks is necessary to do the interesting ones. No one wants to wash labware, redo a procedure for what feels like the thousandth time, or meticulously update their notebook, but all will need to be done correctly if they are responsibilities of the research project you join. **Mentors want to work with students who will push through the uninteresting parts of research without hesitation.** So, when a student shares a version of, "*X* class was really boring and a waste of my time," it's enough for some interviewers to decline to consider inviting the undergrad to join the research group—and it won't matter whether the class was a required course or an elective course unrelated to your major. Also, you have no way to know which classes your potential mentor helped design. When one colleague asked a potential researcher how they liked a lab course, the student launched into a diatribe about how "disorganized and useless" the class was. Our colleague had designed the lab and taught it for several years prior to this interview. Never devalue a class in an interview—even if you felt it was a complete waste of time and tuition money.

» **Sending the message that classes are too much work.** As we previously discussed, for some research experiences labwork or fieldwork is only one of the requirements. Among other tasks, a student might need to read the PI's papers, learn the theory behind techniques, plan experiments, create a poster, or write a report outside of their research hours. Many PIs prefer students who want at least some of these additional opportunities as part of their research experience. Therefore, when an interviewer is told a variation of, "To get an *A* in that class, I had to do everything—go to lecture, take the quizzes, study, and read the book," **it raises the concern that the student might cap their efforts while working on their research project—perhaps midway through a protocol or procedure—if they think it's too much work.** Commenting on how much work certain classes are can also send the message that the student might be more interested in listing research on their CV rather than in investing themselves in a research experience. And, as we've mentioned, many mentors don't offer a research position to someone if they suspect this is the case.

» **Stating that conducting research isn't your actual goal but making up for a classroom deficit is.** You know that the relationship between the classroom and a research experience can be beneficial. However, some students forget that the two complement and reinforce each other, not serve as substitutes for each other. Sometimes, an undergrad will mention using research as a method to reach improbable goals. For instance, a colleague interviewed a student who said, "I didn't do well in my cell biology class, and this is a cell biology lab, so I'm hoping research will fill in the blanks before the MCAT [Medical College Admissions Test]." As their discussions continued, it became evident that the student had planned to do some research but essentially was hoping that their future labmates would serve as private tutors in the subject. **Your research experience can give you an advantage in certain classes and supplement your classroom knowledge, but it won't serve as a substitute for it.**

7. **Broadcasting arrogance.** No matter how confident you are in your research skills, and even if you're an expert in the research techniques that you would use to complete a project, you must establish that you're open to learning. **A student who comes across as arrogant in a research interview rarely receives an offer to join the research group. This is because the student inadvertently sends the message that they will be difficult to teach or work with.** Even if you have more experience than any student in the history of undergrad research, completed a research internship in high school, won an international science fair, have aced all lab classes, and participated in research elsewhere on campus, you should temper your confidence with modesty and demonstrate that one of your goals is to learn.

   Once I (PHG) interviewed an undergrad who insisted that they would be able to design their own experiments by the end of their first research semester because they were "smarter than most people," as demonstrated by their near-perfect score on their college admissions exam. (Our and other mentors' experiences show that standardized test scores are unreliable indicators of a student's future success in a research position.)

   After I explained to the student why their time frame for designing an experiment in our lab was unrealistic, they informed me that I was wrong and would realize it after they started working on a project. I then knew that the student's expectations conflicted with the research experience I was offering, and it was unlikely they would be happy as a member of the research team.

   We're not suggesting that you should minimize your accomplishments—your CV is the perfect place to highlight those. But in the interview, you'll need to emphasize a genuine interest in the research position and indicate that your overall objective is to learn. Using the phrase, "I would love the chance to use those techniques again" or "I want to learn more about that topic" or something similar is a good approach. If you possess an exceptional skill set, after you have the position, you'll demonstrate it by how quickly you meet the objectives of your research project.

8. **Asking the interviewer to solve a *minor* problem.** Research requires both the desire and the ability to solve problems, so avoid asking your interviewer to weigh in on one that you could easily solve yourself by doing a simple internet search or contacting the appropriate administrator on your campus. An example of this from a lecture class would be contacting the professor to ask them when an assignment is due because you didn't feel like checking the syllabus for the answer. Keep your questions relevant to the research position, the interviewer's expectations, and the research program. **However, there is an important exception to this. If you have a serious problem or issue you want to discuss, most inter-**

viewers will help by discussing a potential solution, referring you to the appropriate campus resource, or contacting that resource on your behalf. So, if you're feeling overwhelmed by college, or from managing a personal matter such as a disability or chronic health condition, food or housing insecurity, or another reason, your focus should be on getting the help that you need, not on securing a research position.

9. **Ignoring the interviewer's expectations.** On social media, we regularly receive pleas for help from unhappy undergrad researchers who regret ignoring an expectation at the interview. Typically, the request starts with a phrase similar to "I know my mentor said I that would be doing *X*, but I didn't think it would be this boring/be so hard/take this much time" or other frustration. Our advice is usually to refer the student back to their mentor to try to resolve the issue because we can't alter the parameters of their project or research experience. In most cases, we suspect that the solution for both parties is for the student to search for a new research experience and more carefully evaluate the interviewer's expectations before joining their next project. However, if any of these students would have contacted us before they were in an unfulfilling research experience, we would have shared the following advice with them.

   Interviews are exciting. A mentor's enthusiasm for their research can be contagious. You want to wrap up the search and get started on a research project. So sometimes it's tempting to ignore a potential mentor's requirement or hope that an expectation will be negotiable after you've started a research experience. But this approach to finding a research position could eventually take you down an unhappy path.

   For example, an interviewer might state that the available position is strictly for a paid dishwasher without the possibility of conducting research. Or the opportunity will require you to spend every Friday night in the lab, or the time commitment will be between sixteen and twenty hours per week. **It's up to you to decide whether any expectation or requirement is a deal breaker, an inconvenience, or no issue at all *before* you accept a position.** You aren't obligated to accept a research position if it's not right for you, no matter how nice the interviewer is. As you consider a potential mentor's expectations, remember that nothing will make you more unhappy in the long term than ignoring a significant expectation or a requirement. We assure you, that particular grain of sand won't turn into a pearl.

10. **Not following an interview strategy.** You could wing it by just showing up on time and hoping that the interviewer tells you everything that you need to know, but that approach doesn't put your best interests at the heart of an interview. Following our interview strategy will help you get the most out of those short interactions so you can evaluate the available position

and determine whether it's a good match your goals. Being prepared will also reduce your stress level at the actual interview and help you demonstrate enthusiasm for the opportunity. The rest of this chapter will help you develop your interview strategy.

TIPS SPECIFIC TO VIRTUAL INTERVIEWS

An individual mentor might conduct either an in-person or a virtual interview, whereas most SURP committees do virtual ones. If you're comfortable attending online lectures or giving remote class presentations, a virtual interview won't be more stressful than an in-person one. Although most of the previously mentioned tips apply to all interviews, the ones presented next are specific to online interviews.

- **Choose your space.** Try to find a quiet, distraction-free interview space with dependable Wi-Fi. A good space is a library study room, whereas a bad space would be a coffeehouse or a bus stop. Some video apps include a noise-canceling background feature that can eliminate noises such as typing, but it's typically not a substitute for a very loud space. If you're interviewing at home, do your best to ensure that furry friends, roommates, or family members won't interrupt during the appointment. If this isn't a possibility, then we highly recommend the library study room or contacting your campus Career Resource Center to ask about reserving a similar space.
- **Adjust the room lighting.** Do your best to create a well-lighted space. If you're interviewing in your home space, if needed, place task lights behind your computer and ensure that your face isn't in a shadow. Although it's best avoided, if you must have your back to a window, consider closing the blinds or cover the window with *something* such as a blanket so you aren't backlit. But we do recognize that this approach can be tricky if you need the blinds open for the light *and* your back is to a window. When you do a test of your interview setup with a friend, make adjustments in real time and ask them to help you determine the best solution for the interview environment.
- **Check the equipment.** Interviewers want to connect with you—it's why they opt to have a conversation as part of the selection process. If needed to frame your face and shoulders in the screen, use an adjustable chair to raise yourself up, or put books under your computer to adjust the camera height, or perhaps both. Next, test the sound quality and ease of use of any earbuds, headset, or microphone you plan to use. Although an external clip-on microphone shouldn't be necessary for most people, if you do plan

to use one in an interview, practice using it beforehand. Don't borrow one from a friend the night before and let your first experience using it be during the interview.

- **Pick an interview outfit.** Sometimes students are encouraged to dress up for SURP interviews, which is fine if you already own the clothes. But having to purchase a new outfit for an interview is financially penalizing for many students. And from our informal discussions with colleagues, it probably doesn't make enough of a difference to be worth the cost. But in case the interviewers expect more formal attire, a plain shirt with a collar is a safer bet than a graphic tee. Wear slacks, jeans, or a skirt or dress—whatever you already own that's in relatively good condition. Your shoes can be anything you want, although we suggest skipping the bunny slippers because you might feel less professional.
- **Check, check, check.** A few days before the interview, conduct a test of your setup. By then, you'll know which platform the interview will be held on (Zoom, for example). Whether you're already familiar with the app or need to register for an account, put on your interview outfit and chat with a friend or family member—it can be about your research goals or be a purely social conversation. During this test, ask for feedback on technical aspects with questions such as, "How are the sound, lighting, and camera quality?" and "Is the background noise level low?" and "Is there anything distracting about my outfit, where I'm sitting, or what I'm doing?"

TO SEIZE YOUR UNDERGRAD RESEARCH INTERVIEW, ASK THE QUESTIONS THAT MATTER

Many students approach interviews with a single goal in mind: get an offer to join the research group. Although landing an offer to join a group is an appropriate goal, it shouldn't be your *sole* objective of interviewing. Obviously, you want to highlight how you'll be an asset to the research team, but it's equally important to learn as much as you can about the research project or position, your potential mentor's expectations, and areas of lab culture most relevant to your experience.

Essentially, your goal is to know the answers to the following self-questions by the end of an interview:

- Am I genuinely interested in working on the available project or joining the group in a nonresearch position to start?
- Can I meet the required time commitment without compromising my academics or well-being?

- What are the basic requirements of the position and the interviewer's expectations for how I will fulfill those requirements?
- Will I likely be able to accomplish my goals with the available research position? Or if the opportunity is a nonresearch position, what metrics will be used to determine my eligibility for moving into a research position?
- Does the available research project or opportunity align with my values?

Therefore, as soon as you send out First Contact emails, continue to use the hours you scheduled for a research experience to prepare for interviews by thinking about the goals you wish to achieve through that research experience and preparing questions to ask yourself and your interviewer. The next sections will help you do both.

STEP 1: REEXAMINE YOUR EXPECTATIONS AND WHAT YOU WANT FROM A SPECIFIC RESEARCH EXPERIENCE

Even if you applied for a research position in response to an advertisement, it's unlikely that all the mentor's expectations and requirements will be addressed until the interview. Read the questions that follow and carefully consider how you feel about each in case they are asked by the interviewer. In addition to preparing you for the interview, this exercise will help you identify your deal breakers (if you have any) and clarify *your* expectations for a research position. **In the list presented next, highlight any questions that are important to you or will be an issue if it's required for a position.**

- What do I hope to gain from a research experience? Do I have one goal that is essential? Are some goals flexible if otherwise the opportunity is something I want to do?
- Why am I interested in *this* research group or project?
- Would I be willing to wash lab dishes or join the lab as an observer if I'm not offered a project to work on right away?
- Will I only be able to accept a research or general lab assistant position if it's a paid position?
- How many hours each week am I able to commit to a research experience? What is my maximum number of hours per week?
- What, if any, adjustments to my schedule can I make if my ideal research schedule won't work for a mentor?
- If interviewing before the drop-and-add period ends, will I be willing to change my class schedule to accommodate a mentor's training schedule? What if it means registering for an 8:00 a.m. computational biology class?
- Ideally, do I want a short-term (a semester) experience, a longer one (at least a year), or to continue from now until I graduate?

- If given the option, do I want to start this semester or next?
- Will I be available to continue a project in the upcoming summer? Will I be willing to increase my weekly hours during the summer, if asked? What is my summer weekly hour maximum? Will I only be able to conduct research in the summer months if the research position becomes a paid opportunity?
- Which, if any, academic breaks am I willing to spend working in the lab or at a field site?
- What do I anticipate will be the most challenging part of undergrad research?
- If registering for course credit is a requirement, or is not allowed, will either stipulation create a problem for me?

If you're interviewing for a SURP, you might have already learned most of the previous answers on the program's website that this section is intended to discover. But you might also need to think about the following questions:

- How do I feel about living away from my family for an entire summer? What if they are unable to visit or I'm unable to travel home for that time?
- Will I need to sublease my apartment (if allowed) or does my lease end before the program starts? Will I need to search for a new apartment to return to in the fall while I'm still at the SURP location?

STEP 2: PREPARE A LIST OF QUESTIONS ABOUT THE POSITION AND THE RESEARCH PROJECT

Preparing questions ahead of an interview will empower you to ask informative questions to determine whether the project, training opportunities, and research experience constitute the right fit for you. Plus, undergrads who arrive at a research interview with appropriate questions appear to be more prepared, professional, and enthusiastic about the opportunity than undergrads who don't. In addition, to get the most out of your interview, **you'll need to ask questions if the interviewer doesn't cover important information. That might seem easy (or obvious), but if you haven't been part of a research experience, how do you know what questions will give you the most meaningful information?** Many interviews are short and therefore don't provide the luxury of time to ask low-value questions.

Take for example, questions such as, "What equipment does your lab have?" or "How many grad students are in your research group?" Answers to those questions will give specific information about the research group, but—and here is the key—what will you do with that information? How will you use it to evaluate the position or to decide between two or more research positions if you receive multiple offers?

First consider the question about lab equipment. If the interviewer rattles off a list, chances are you won't know what most of the equipment is or what is needed for the available research project. In this case, asking the question wouldn't be helpful. However, if you have an independent research project in mind and you know you'll need specific equipment, then asking would be essential.

As for personnel, **it's nice to learn how many people are part of the research group, and in what positions, but it's only information—not a meaningful metric to evaluate an undergrad research position.** For example, if an interviewer says, "Zero undergrads, three postdocs and two grad students," or "Two professional researchers, two undergrads, and four grad students," how will you compare the two research groups? How will you evaluate which is the better choice for you?

Is it better to be in a research group with several postdocs and be the only undergrad student? Would you receive extra mentoring, or would the postdocs devalue your contributions because you're "only" an undergrad?

Or is it better to be in a research group with several grad students and but no professional researchers? Does that indicate a professor who places a higher value on mentoring students over training professional researchers? What about all-undergrad labs? What if the PI mentors only a few students at a time or has more than ten? How do use you that information to determine whether the PI embraces mentoring or is using undergrads as so-called free labor?

And finally, what if a professor is just establishing their research program and you would be the first member? Would you have the opportunity to help set up a lab and receive significant personal instruction and mentoring, or would it prevent you from getting much research done because you'll be busy putting items in cabinets and on shelves?

By asking about personnel in an interview, you'll learn who's part of the research group and in what positions. However, without actually working with those individuals, you can't know how they work together and how that will affect your research experience. Any opinion you receive from someone about how to evaluate a research group based solely on its personnel will be influenced by their personal baggage and won't necessarily reflect the realities of the one that you interview with.

We're not suggesting that you shouldn't ask about who is a member of the research group—but to keep in mind that the information might not be helpful when it's time to evaluate the potential research experience. And although not all PIs regularly update their website, a visit to the People or Lab Team tab might give you some, if not all, of this information.

However, **there are several questions that will provide information on the research experience and the mentor's expectations that will help you carefully consider the opportunity.** For instance, you want to determine whether the research experience will help you accomplish your long-term goals. If pursuing an MD-PhD or graduate work in STEMM is in your future, you might want to ask whether you'll have the opportunity to work on an independent research project after you've been involved in research for a while. Likewise, if you eventually want your research mentor to write a recommendation letter that covers your ability to work well with others, confirm that the majority of your work will be done in a shared work space, not in a room down the hall or in another building.

Essentially, **your preinterview question strategy is this: imagine that the first thing the interviewer will say to you is, "What questions do you have for me?"** before they have explained anything about the project or mentioned their expectations. This will help you determine the essential information you need to glean from every interview. To help prepare interview questions that are important to you, reread the chapter 2 section "Lab Culture" and the chapter 3 section "Understanding and Managing Your Expectations."

In most interviews, you won't need to ask all the questions that are listed next because most mentors will cover these basic details. Also, not all these questions will be applicable to your situation.

Questions about the Time Commitment (Relevant to All Positions)

- How many hours per week are required?
- What blocks of time are required?
- How many days per week, or which days per week, are required?
- How many semesters make up the total commitment?
- When is the start date? When is the end date?
- Will I coordinate my schedule with you or another group member?
- Does the position require conducting research during academic breaks such as spring break? (Or ask if working on your project during academic breaks is an option, but only *if* that is something you're interested in doing.)

Questions about a Research Position

- Is this a paid, academic credit, or volunteer position?
- If it's an unpaid position: Is there a possibility that it can become a paid one?

   » If yes: What is the timeline for that happening and what metrics will be used to determine if I'm eligible for a paid position?
   » If no: Can I start earning GPA credit during the first semester instead of volunteering?
- Is this primarily an on-campus or virtual position?
- Is it possible to continue working on the project until I graduate? (Ask this question only if you *might* want this option.)
- Is there a research contract or learning contract required by the PI, mentor, or department?
   » If yes: Do you have additional requirements or expectations that aren't listed on the contract?

Questions about Virtual-Only Projects

- How often will we meet to discuss my project?
- Will we meet virtually or in person? (You might have a preference for one or the other. For example, if your internet service is unreliable, you might wish to meet in person. If you're managing a chronic illness or have unreliable transportation, or are mostly enrolled in an online degree program, virtual meetings might be the better option for you.)
- Will I contact you or someone else if in between those meetings I have questions?
- What is the best way to contact you (or the person who will supervise my project)?

Questions about a Research Project

- What is the title of the project I will work on?
- What overall question or problem does the project address, and why is it important?
- What techniques will I use at the start of the project?
- What are some keywords associated with the project? (You'll use these later to prepare for your first day of research.)
- Does the lab maintain an archive of published papers that I should be aware of?
- Will I need specialized software for my research project?
   » If yes: What is the name of the software?
   » Will the software be provided to me free of charge?
   » Can you recommend any internet forums for learning and working with the required software?

Questions about Your Immediate Supervisor (If That Person Doesn't Conduct the Interview)

- Who will be my direct research supervisor?
- Are they available so I could meet them today?
  - » If no: Can you please spell their name for me? (Write this down! You'll want to email them to introduce yourself later if you're offered and then accept the position.)

Questions for Your Immediate Supervisor (If They Are also the Interviewer)

- How often will we meet to discuss my research project or responsibilities?
- Will we meet in person or virtually?
- What is the best method to contact you if I have a question and you're not in the lab?
- Is there anything I can do or read before my first research day to prepare?

Questions about Accommodations for Students with Disabilities

- Does the current project or work environment have the appropriate accommodations in place that I need? (You might or might not be comfortable opening a discussion about the accommodations specific to your situation with the interviewer, and they might or might not know enough to answer your question right away. Instead, they may need to work with you, and possibly a campus official, to put a plan in place, as covered in the next question.)
- Will you work with the Services for Students with Disabilities Office to put reasonable accommodations in place? (You can also make this request after joining a research experience even if you didn't feel the need for accommodations when you interviewed, or you didn't feel comfortable discussing the matter at the interview, or if your situation has changed since the interview.)

Questions about Safety Training and Ethics

- Will I need training in lab safety, animal handling, or fieldwork procedures?
  - » If yes: Do I complete any of these before I start working on the project?
- Are vaccinations required before I start? Which ones?
- Will I be issued keys or a pass code to access the building or labs?
- Are seminars or workshops covering ethical research required before or after I start?

Questions about Research for GPA Credit (Customize to Your Situation)

- Can I register for course credit the first semester of research? (Or register for a 0-credit research option if it's available *and* you're interested in it.)
- Is there a research contract required by the department or honors department to receive GPA credit for conducting undergrad research?

Questions about Writing Papers, Reports, and Presentations

- Will I write a research preproposal or end-of-semester report (or both)?
- Will I give a poster presentation or short talk at the end of the semester either to the research group or at an undergrad symposium?
- Is earning a publication coauthorship a possibility with the available project?
- Could I write an undergrad thesis or an undergrad research paper on the project? (Ask only if you're genuinely interested in these possibilities.)

Questions about a General Lab Assistant Position

- Is this a paid position?
  - » If yes: What is the hourly rate? Is there a maximum number of hours I can work this semester? Is a work-study award required for this to be a paid position?
  - » If no: Is there a possibility that it can become a paid position?
- What are the core responsibilities of this position?
- Is there a possibility of switching to a research position after working as a general lab assistant?
  - » If yes: What is the timetable for this happening?
  - » How will you determine whether I'm eligible to move on to a research position?
  - » Will I still be paid the same hourly rate after becoming a researcher? (Note: You probably cannot earn course credit and a paycheck at the same time.)

Questions about an Observer Position

- How long do students typically observe group members before starting a research project?
- What could students who weren't offered a research position after that time have done differently?
- Will I shadow the same labmate each session or will it change?
- Do I introduce myself to the research group or do you?

STEP 3: REEVALUATE YOUR PROPOSED RESEARCH SCHEDULE

At a research interview, it's essential to answer questions about your schedule with honesty and confidence. If you've been testing your academic and life balance with the hours you set aside for research at the start of chapter 4, the next part will easy. Ask yourself these questions: "Did the hours I scheduled for research work for me?" and "Is my maximum number of hours per week still the same?" If adjustments are needed, review the information on scheduling in chapter 4 and revise your schedule.

STEP 4: PREPARE TO DISCUSS ADDITIONAL TOPICS

By following the search strategy from chapter 4 and reading this chapter, you're already prepared for most research interview questions. But you might not be able to anticipate everything in advance because interviewers will ask questions about their specific research opportunity. If you're surprised by a question at the interview and need a minute before answering, that's okay. Simply answer, "Wow, good question. Let me think about that." Then take a moment to gather your thoughts before answering. Depending on the interviewer's style, there might be a few other topics that come up. How to prepare for them before the interview is covered next.

Questions about an Advertised Project

Advertisements for research positions are an excellent resource to help you prepare before an interview. If you're fortunate to have access to one, read the advertisement thoroughly and try to learn the meaning of every word and concept you don't know. This information will help you answer questions similar to, "What inspires you about the available project?" or "What do you think this project is about?" which are reasonable to expect in an interview, even if you already covered them in First Contact (as described in chapter 5). Here again, it's not expected that you'll be an expert in the project or the science the lab does. But if you gained a basic understanding of what question the study addresses or why it's important, that bit of knowledge helps demonstrate enthusiasm for the project.

Questions about Your Lab Classes

If your interviewer teaches a section of a lab class you've taken, or was involved in designing it, they might have a set of questions based on the course. They might want to know if the letter grade on your transcript reflects the knowledge you gained from the class, or how interested you were in the subject material. Or they might just bring up the class as a way of making conversation to break the ice. It's unlikely that they will ask you about every procedure done in the class—probably just one or two at the most. Prepare to discuss which lab classes you liked and why, or to share details about an experiment or technique from your current lab class. *This is especially important if you referenced a lab class as your inspiration for applying to a research project.*

Questions about a Technique You Emphasized in First Contact

If you mentioned a specific technique in First Contact or on the application, be prepared to discuss it. You don't need to know every detail about the technique, but you should know something about how it's carried out, the reason it's used in research, and why it inspired you to find a research experience that includes it. Stating, for example, that you loved the electrophoresis unit in your general biology lab class because you're fascinated with *E. coli* probably won't convince the interviewer that you know much about the technique that supposedly motivated you to apply to their research program.

The Dreaded Question: "What Is Your Greatest Weakness?"

Not all interviewers ask candidates to explain a personal weakness, but some do. At a research interview, when an interviewer asks you to explain your greatest weakness, generally it's an attempt to assess whether you might be difficult to work with, or whether you truly value the experience you're interviewing for, or whether you have an issue with time management that might have a negative impact on your training. Many students answer this question with a variation of "I'm not sure how I'll be able to stay on top of my coursework and do research." We've already discussed why the absence of time management skills isn't a reassuring answer to a potential mentor and actions to take to help avoid this scenario.

Although there is no one single right answer for the "weakness question," we suggest something along the lines of, "I don't have any research experience." And if the interviewer asks the obvious follow-up question, "Why is that is a weakness?" respond with, "Well, I don't know what I don't know. But I'm open to learning it all."

But if you have conducted research in a professional lab and you're interviewing for a new position, your response phrase might be more like, "My previous project was on microbiology, but this project is about population ecology. So, I don't know what I don't know, but I'm open to learning it all."

STEP 5: SCHEDULING THE INTERVIEW

Whether your interview will be held in person or virtually, most likely you'll schedule it through email. So, keep the following tips in mind during the scheduling process:

- **Respond to an interview invitation within twenty-four hours of receiving it.** You want to demonstrate enthusiasm for the position and show that you have time to conduct research. Waiting a few days to respond can send the opposite message to some interviewers. Also, if the number of interview time slots is limited and other candidates respond quickly, a lag in your response could cost you a slot. We know that a twenty-four-hour turnaround will be difficult for some students, so do the best you can. One additional tip on this topic: regularly check your email spam folder in case an interview invitation is accidentally delivered there.
- **Consider your commute before proposing, or agreeing to, an interview time.** If you have to rush across campus because a class went long or you needed to connect with the professor after, or a bus was late, a last-minute dash might make you more stressed out than if you had a time buffer. When scheduling the interview after a class or a work shift, give yourself enough time to review your notes and still arrive at the interview five to ten minutes early. Even if your interview is a virtual one, you need time to commute to the interview space and set up the device you'll use for the interview. (There are more details on reviewing your notes before an interview in the subsequent section "Preparing for the Interview.")
- **If the interviewer uses an online app for scheduling, follow all the instructions.** For instance, by default some apps allow the selection of multiple appointment times, but if the interviewer instructs you to choose only one, do that. Don't select multiple times unless the interviewer requests that you do.
- **Make it easy for the interviewer to schedule time to meet.** If they don't use an online scheduling app and ask for times that you're available to meet, be absolutely clear to reduce back-and-forth emailing. Stating "I'm available Tuesday afternoon *at 2:00 p.m.*" is more precise than "I'm available on Tuesday *after 2:00 p.m.*" The exact time given in the first example is easy to immediately schedule, while the second example leaves the interviewer wondering, "Can they meet at 2:05? or 2:15? or do they actually mean 2:30?"

- **Once scheduled, avoid changing the interview time.** Yes, sometimes conflicts unexpectedly arise, but make sure it's worth risking your interview. If you want to change the appointment time, you might not be offered the option to reschedule. At any given time, most researchers have a bloated schedule and several pending appointments that make scheduling anything overly complicated. If they have already scheduled interviews with other undergrad candidates, a potential mentor might complete those first and only reschedule an interview with you if none of those work out. This isn't fair, we recognize, and we believe that most mentors will try their best to reschedule, but sometimes it doesn't happen.
- **Schedule interviews close together.** If you have multiple interviews, try to schedule them within a day or two of each other. You'll want to decide quickly if you're offered multiple research positions. This might not be possible, and it might not matter if you interview with the first lab and accept the position as soon as it's offered, but scheduling the interviews only a day or two apart is worth trying to arrange if you can.
- **Give yourself plenty of time for the interview.** Ideally, the interviewer will mention how long they anticipate the interview will last when they ask you to schedule a time for it. However, if they don't, avoid asking for an estimate unless you really need one to ensure that you make it to a class, a job, or another appointment. Although some interviewers will interpret your asking as a sign of organization, others will interpret it as a sign of an already overcommitted student trying to squeeze an interview in between more important activities. From our conversations with colleagues, this difference in interpretation tends to be influenced by a mentor's most common experiences working with undergrad researchers and not what they suspect about a specific person. Try to schedule your interview with enough time to accommodate the possibilities covered next.
  - **Choose an interview slot after your classes and obligations are done for the day or when you have at least sixty to ninety minutes of unscheduled time.** You don't want to be nervously staring at the clock or wondering if you'll be late to your next appointment if the interview goes long. If you need to cut the interviewer off to leave, or broadcast that you need them to wrap up it up for any reason, you'll send the message that you have somewhere more important to be. This will generally be interpreted that you're either overcommitted or uninterested in the research opportunity. And although we don't anticipate that an interview will last as long as ninety minutes, you might need time to commute to your next activity or grab a snack before getting there.
  - **If you don't have anything scheduled directly after the interview, and you're offered the position, you might have the opportunity to**

get started that day. Some PIs like to bring a new student from their office into the lab to make introductions to the other team members. If the person you're going to directly work with is around, you might have the opportunity to ask what you can do to prepare before your first day or simply have a social conversation. Also, you may be able to observe a technique that your in-lab mentor or another labmate is doing. At the very least, even if you're only introduced to your in-lab mentor, it breaks the ice. About half the students we offer a research position to start in the lab the same semester they interview. Many begin by meeting their labmates right after the interview, although they don't start benchwork until after they have completed their safety training.

MENTORS' INTERVIEWING STYLES

When interviewing for a research position, the interviewer might be the PI, a grad student, a postdoc, or a staff scientist. If you're interviewed by the PI, it could be to work directly with them or to assist another member of the research team. If you're interviewed by someone other than the PI, you'll most likely work directly with that person—possibly on a subproject created from one of their projects or on a project that you propose.

As with applications, there isn't a standard interview format for undergrad research positions. If the available opportunity is an observational one, formal interviews tend to be short because most of the interview takes place as you're interacting with the other members of the research group. If the available position is for a general lab assistant, or a researcher with an independent project, interviews tend to be on the longer side, although most are still relatively short.

Each interviewer selects the format that works best for them. So, if you compare research interviews with a classmate, the interview experiences might be drastically different. For example, if you're asked several intense questions about the research topic but your classmate mostly is asked about their hobbies, chalk it up to different interviewing styles.

Also, some interviewers use a curated set of questions to interview all potential researchers and others just wing it in the moment. One interviewer might ask several questions about your instructional lab experience, transcript, or extracurricular activities, while another interviewer won't ask any questions about anything and will basically chat with you about various topics. When it comes to answering questions, some interviewers are patient and will address all concerns a student has even if it makes the meeting longer than they had initially scheduled. Other interviewers stick to a predetermined time allowance and will begin wrapping up an interview when they approach it because they have something else scheduled right after the interview.

If a specific project is available, most interviewers will explain an overview of why the project is important. They will also state the position's requirements: *You'll do A, B, and C, and you'll work with D. The time commitment will be X hours a week for Y semesters.*

Ultimately, each interviewer wants to determine whether you're genuinely interested in the research, whether you're able to uphold the required time commitment, and whether they recognize any issues that would likely be problematic after you join the research group. When relevant to the position, some interviewers will also want to determine whether your schedule is compatible with your in-lab mentor's or if you have a specific skill set or other prerequisites.

Next are generalized examples of four interview styles based on our approach (which most closely mirrors Style 2), and those of colleagues who shared their styles with us.

INTERVIEW STYLE 1

The interviewer uses highly technical language to describe their research program and potential projects. They might explain one or two available projects and then ask which you would like to work on. Alternatively, they might tell you that Researcher A works on Project X and Researcher B works on Project Y, so you should meet with both individuals to determine which is more inspiring to you. This can be a little intimidating to do—setting up appointments to meet with two additional people you don't know to ask about their projects—but the people in research groups that conduct these types of interviews are usually quite friendly and excited to share details about their project and how you would contribute to it. For tips on what questions to ask each researcher about their project, revisit the chapter 4 subsection "Attend a Research Symposium on Campus," within the section "More Methods to Identify Undergrad Research Opportunities."

INTERVIEW STYLE 2

The interviewer starts by giving an overview of their research program and the available project in nontechnical language and then covers their basic expectations for undergrad researchers. But much of the interview is spent asking you questions about your CV, transcript, experiences you've had in classes, or what you want to accomplish through a research experience. The interviewer may also talk about their career, why they became a researcher, the importance of mentoring undergrads, or similarly related topics. You know it's an interview, but it feels more like a conversation you'd have with a friend because the atmosphere is so casual.

INTERVIEW STYLE 3

The interviewer discusses their research program or the available project in detail using lots of technical language or nontechnical language. They ask you some general questions (why you want to do research, what your long-term career goals are) and examine your CV and transcript. **If they believe that you're genuinely interested in their research, they assign you several papers to read** related to their research and instruct you to follow up after doing so. If you return to discuss the papers, and do well, you'll be invited to join the research group.

INTERVIEW STYLE 4

The interviewer goes over an available project quickly, in either technical or nontechnical language. **Most of the interview is spent discussing your schedule**, and you're offered the position without being told what the interviewer's expectations are (in which case, you'll need to ask). Unless an unexpected issue with your answers or enthusiasm pops up, this type of interview is mostly a formality, as the decision to offer you the position was made at First Contact.

PREPARING FOR THE INTERVIEW

If you've followed the strategies suggested in this book so far, you're almost prepared for your research interview, but there are a few easy tasks to complete before the interview day arrives—some of them the day before and a few ten minutes before the interview.

ONE DAY BEFORE THE INTERVIEW

Life has a way of throwing minor complications in your way—your roommate is loud all night, making it impossible to sleep; a surprise fire drill occurs; your morning alarm doesn't go off—that can make it hard to focus on priorities at the last minute. **Complete these tasks the day before your interview.**

1. **Be ready to share essential documents with the interviewer.** Whether you've already emailed these to the interviewer, it's a good idea to have backup copies ready to go.[2] Therefore, print, upload into a sharable folder in the cloud, or keep in a draft email that you could send to the interviewer in the room, the following:
   » Anything the interviewer has instructed you to bring, such as proof of a work-study or other financial aid award
   » The schedule you prepared during your search
   » Your transcript and CV
   » A list of the classes you intend to take next semester (just what you plan to register for, not the sections or class times, which you may not yet know)

   And print or email to yourself:
   » A copy of the questions you want to ask the interviewer
2. **Determine how you'll take notes.** If you plan to use pen and paper, put them in your backpack; include plenty of paper and at least one extra pen in case the first one breaks or runs out of ink. If you plan to take notes on a computer or other electronic device, charge the device the night before.
3. **Select your interview outfit.** You're a student, so it's fine to dress like one—whatever that means to you. For example, G-rated graphic tees are fine, as are jeans and athletic shoes. If you want to dress up because it makes you feel confident, that's okay, too, but it's not required, and we don't recommend purchasing clothes for this occasion. If you've taken a lab class that outlined specific dress requirements for the type of research that the group you're interviewing with conducts, use those rules to choose your interview outfit. If you haven't completed a relevant instructional lab, good practices include wearing closed-toe shoes with a stable bottom, choosing eyeglasses instead of contact lenses, and selecting slacks instead of shorts.
4. **Review the interview mistakes to avoid.** Sometime during the day before the interview, do a thorough review of the section "Ten Topics to Help Avoid Interview Mistakes" at the beginning of this chapter. But don't do this review right before bedtime because it might stress you out or prevent you from sleeping well.

2. Although we're pursuing a minimalist approach to printing documents in our lives, we recognize that not all students will interview in rooms where the Wi-Fi is accessible, so you may wish to print these documents before the interview. If you don't own a printer, check with your Student Services Office or Study and Writing Center for access to one. Many colleges allow students to print a limited number of free pages each semester.

TEN MINUTES BEFORE THE INTERVIEW

To put your brain into "interview mode," about ten minutes before the interview, reread some basic information about the position. You don't need to reread everything connected to the research opportunity, but choose one or two pieces of information. Consider, for example, reviewing your Impact Statement, the advertisement for the research position, the application questions and answers, or the email correspondence you exchanged with the interviewer. You could also go over the questions that you want to ask the interviewer. Remember, *it's okay to be nervous, and most people are at interviews*. Providing your nerves don't prevent you from answering questions or asking for clarification when needed, consider being nervous a little bit of healthy excitement.

Good luck! If you've done what was outlined in this chapter, you're ready!

AT THE INTERVIEW

You'll go on many interviews during your career—maybe as a candidate for grad, med, or professional school, as a potential employee, or as an entrepreneur trying to secure a bank loan to start a small business. Although you won't ask the same questions in these types of interviews as you would for a research position, you can use the preparation work and the self-reflection strategies covered in this section to help you in those types of interviews as well.

TAKE CONTROL

After you meet the interviewer, there might be an awkward pause before the interview begins. It's not crucial, but if you feel like doing it, this could is the time to say, "I'm excited to learn more about __" and fill in the blank with "the available project," or "the techniques your lab does," or "your research program." It's a simple thing to do, but it's not an obvious one and it's another way to share that you're eager to learn.

TAKE NOTES

It doesn't matter if you take notes on paper or an electronic device as long as you get the important information down. However, if you do take notes on your computer, phone, or tablet, inform the interviewer before you start typing. That way, they won't wonder if you've checked out during the interview or are more interested in updating your status on social media than learning from them. A quick statement, "I'm going to take notes on my computer," will suffice.

There are **two basic but important reasons to take notes:**

1. **To refer to the details later.** Even if you accept the research position at the interview, you might not start on your project for several weeks. Relying on your memory for something as important as your mentor's expectations is a risky strategy because most mentors don't plan to go over their expectations again after you start working on your project. Writing down those expectations during the interview will provide a reference on how to get the most out of your research experience and earn the strongest recommendation letters.
2. **To evaluate the position.** Interviews are exciting, and it's easy to get caught up in the competition of winning the position. Your notes will help you evaluate whether the experience will be a meaningful use of your time, or they can become the basis of a pros and cons list if you need to decide between two positions. The notes you take on the professional development opportunities and relevant areas of lab culture will be particularly helpful in this regard.

ASK QUESTIONS

As the interviewer starts to wrap up the conversation, they will probably ask you a variation of "What questions do you have?" For you this is an important part of the interview process. Whether because you're worried that the interviewer won't think that you're smart or you're afraid the answer might not be what you want it to be, not asking a question is a poor interview strategy because **not asking a question won't change the answer—it will only make you less informed.** However, asking a question because you're genuinely interested in more information helps you evaluate the opportunity and demonstrate that you're enthusiastic about it.

Even if the interviewer answered all your questions during the interview, **confirm or ask about the time commitment** to ensure that you're absolutely clear on their expectations. The actual question you ask depends on what information already has been given by the interviewer. Consider the following two scenarios to determine your question.

Scenario 1: The Interviewer Has Not Told You the Required Weekly Hour Commitment

If the time requirements haven't been addressed, your first question should be, "*So I can meet your expectations*, can you tell me how many hours per week I will work on the research project, and in what blocks of time?" As the interviewer responds, take notes. If you're still interested in the position, and you know you can accommodate the research schedule, tell them, "This is good. I've already worked out my schedule, so I know I can do that." If you can't meet the time requirements, leave off the last sentence and simply nod to indicate that you understand the requirement. (You'll learn how to handle this possibility later.)

Scenario 2: The Interviewer Has Told You the Required Weekly Hour Commitment, and Your Schedule Is Compatible, and You're Still Interested in the Position

Consult your notes for accuracy, then repeat back the required time commitment. Start with, *"I'd like to make sure that I understand your expectations,"* and then fill the rest in with the previously mentioned requirements. When the interviewer confirms that you're correct, your next statement should be, "This is good. I've already worked out my schedule, so I know I can do that."

After you've demonstrated that you understand the expected time requirements, ask the rest of your questions or for clarification on any topic that was brought up or occurred to you during the interview.

What to Do if a Deal Breaker Is Mentioned

Sometimes you won't know the all the requirements of a research position until you learn about them at the interview. For SURPs this isn't the case because each program covers their expectations on their website. But **if the interviewer mentions a requirement that is a deal breaker for you, the ideal time to address it is after you've been offered the position.** This might be before you leave the interview room or after you receive an invitation to join the research project through email. **However, if you do bring up a deal breaker, you must do it carefully.** Don't make the mistake of trying to explain to the interviewer why they are incorrect about requiring your deal breaker because it won't help you.[3] A better approach is to respectfully decline the offer and state the reason without asking for special consideration. This approach lessens the chance that you'll offend the interviewer but still makes it clear that you'd like the position *if the requirement is actually flexible*. This also requires the interviewer consider what is

3. *Deal breakers* in this section don't refer to institutional barriers that prevent a student with disabilities from having appropriate accommodations in place, which are your right to have, not an option for the interviewer to decide whether to provide.

more important to them—getting you involved in the research project or maintaining the requirement.

Here's how to do it: You'll want to decline the offer with a strong statement that reflects something you learned about the project, research group, or techniques in the interview. For example, say, "I am so excited about this research project because I like the idea of using PCR to determine the chromosomal location of a gene of unknown function. However, I have only twelve hours to dedicate to research per week—any more and I'll risk becoming overcommitted. If eighteen hours per week is firm, I will, unfortunately, need to decline the offer to join the lab."

At this point, the decision is in the interviewer's court, and they might be open to a solution that works for you both. However, they might respond that there is no flexibility on the issue because the requirement is a deal breaker for *them* or the *position*. Therefore, remember that if you use this strategy, you must be prepared to lose the position. **Once you declare that something is a deal breaker, there is no possibility to take it back.** Be certain that your reason is something that will prevent you from keeping your academic and life balance intact or that will make you unhappy or risk your well-being—not an inconvenience that you're hoping to negotiate.

AT THE END OF THE INTERVIEW

If you're offered the position at the interview, and you believe it's the right one, accept it immediately. If you're unsure, tell the interviewer that you'll need to think about it. If you don't want the position, you can inform the interviewer or email them later with your decision. If you want the position but cannot accept it because of a deal breaker, you'll want to address it as described previously.

Most interviewers won't want you to feel pressured to accept a position, so if you aren't offered the opportunity at the end of the interview, you might need to make it clear that you're interested. **As the interview draws to a close, ask, "When will I know if I get the position?"** You can also say to the interviewer, "I hope to work with you because ____." And fill in the blank with something specific you learned at the interview or why you're excited to get started. This type of statement leaves a strong impression and will distinguish you from the other candidates who don't include a similar statement at the end.

AFTER THE INTERVIEW

It's important to thoroughly evaluate the research opportunity while the interview experience is still fresh. Remember, your goal isn't simply to get a research position but to choose an opportunity that is a meaningful use of your time. It doesn't matter if you accepted the position at the interview or if the interviewer doesn't contact you for a few days. You still need to evaluate how you truly feel about the opportunity before your memory of the interview fades.

In addition, if the interviewer offers you the position, you'll want to respond with your acceptance (if you want the position) within forty-eight hours, but sooner is always better. The fast turnaround can matter because once a position is offered, the clock starts ticking. If you take too long to accept, some interviewers will presume that you're not interested but are afraid to say so. In these cases, they might offer the position to another student so they can wrap up their search. Other interviewers won't be in a hurry. The trouble is, you won't know how long a particular interviewer will wait before offering the position to another student unless they inform you. And if your offer is for a SURP, they may or may not allow flexibility in the response deadline. Therefore, if you evaluate the position shortly after your interview, you'll be prepared to respond to the interviewer quickly in most, though not all, cases.

EVALUATE THE OPPORTUNITY

Start the evaluation process by thinking about how you felt when the interview ended. Were you inspired and couldn't wait to get started? Alternatively, were you bored during the interview and glad when it was over so you could leave? *How you felt at the end of the interview is a good indication of how you feel about the opportunity overall.*

The next step is to ask yourself the following:

- Am I genuinely interested in the techniques, project, or subject?
- Can I work the required hours without compromising my academics? If I work at a job, can this opportunity coexist with that schedule or offer class credit or payment for participation?
- Do I *want* to make the required time commitment in hours per week and semesters?
- Do I want to join this research group? Do I want to work with the person who interviewed me? Did the interviewer say or do something that makes me reluctant to work with them?

- Am I tempted to accept the position so I can quit searching for one even though I'm not very interested about the opportunity or because the interviewer was nice and I don't want to hurt their feelings?
- Will I likely be able to accomplish my goals and gain the interpersonal and professional development I want through the available position?

Think about Your Response

What happens next is related to the outcome of the interview and how you feel, as presented in the following possible questions you may ask yourself, depending on your own situation.

What should I do if I'm still unsure about the position?

If you're still ambivalent after evaluating the opportunity as described, try to determine why. Is the required schedule a problem or the type of research different from what you want to do? Was the interviewer intimidating? Are you worried that the project seems too complicated and you're afraid of failing? Review the relevant categories under the chapter 2 section "Lab Culture" and the chapter 3 section "Understanding Your Expectations" to help identify anything that might explain your ambivalence. **If your hesitation is because of an overall feeling of nervousness about starting a research experience, know that you're in good company—most students feel the same way at the start.** But it would be a shame if you turned down a research position because the unknown was intimidating.

If your hesitation is for a specific reason—such as the project or techniques don't seem like something you want to do, or there is a requirement that you cannot (or don't want to) fulfill, or the type of experiential learning offered isn't what you want—**then it's best to decline the position and keep searching for something that aligns with your goals.** If you still can't decide, keep in mind that indecision itself is often a decision. It's okay to think it over, but if you aren't sure after three days, it's probably not the right project or research group for you.

What if I'm offered the position and it's the one I want?

This is easy. Respond immediately with, "I can't wait to get started. Thank you!" If the offer comes to you through email, include your proposed research schedule in your reply to ensure that you and the interviewer have the same understanding of what it is.

What if I receive an offer, but I have another interview scheduled?

This can be tricky. On the one hand, you want to explore all your options and make a choice based on the most information. On the other hand, research positions are competitive, and you can't be sure how long an interviewer will hold the position for you after making an offer. (This is why, ideally, you want to schedule interviews within a day or two of one another.) If you determined that you would like to join the research group during your postinterview evaluation, you could go ahead and accept the offer. What you shouldn't do, however, is accept an offer and then continue to go on other interviews or ignore an email with an offer.

Also, some SURPs have short deadlines for accepting or declining a spot. If this is your situation, we recommend connecting with a science professor or mentor for customized advice on how to proceed.

What if I receive two offers?

Of all the tough decisions to make in life, this is a pretty good one. If both opportunities seem like ones you'd like, you'll need to use other criteria to choose one. **Examine your schedule to determine whether one position is more compatible with your academic and life balance. Then review the overall expectations and the answers to the questions you asked at each interview.** For instance, does one offer a paid position while the other is volunteer only? Next, if you're still unsure, use the internet to learn more about the techniques you'll be using in each project and reread each PI's research interests. Finish by considering what you know about the lab culture, paying special attention to the training opportunities and your ultimate goals. This advice is all relevant to SURPs as well, but in those cases, you also might want to compare the financial packages and the location the research takes place. One program might immediately become the clear winner. If both opportunities still seem equally perfect, flip a coin. It's likely that each one is exciting, which is why it's so hard to choose.

What if I don't want the position (or I accept the position and realize before I start that it's not the right one for me)?

If it's not the right experience for you, it's not the right experience. Although it might feel awkward, the best way to get through turning down a research position is to do it quickly and professionally so the interviewer can offer the opportunity to someone who is interested. Therefore, on the same day you receive the offer, or on the day you realize that accepting the position was a mistake, send a short email to decline the offer. The interviewer might be disappointed because they were hoping to work with you, but they won't be upset with you. However, if they are a rare case and are rude to you, know that you absolutely made the right decision by choosing to not work with them. There is never an advantage to accepting a research position that you don't want just to make someone else happy. And no matter how nice the interviewer was, you don't owe them your time. And you don't need to apologize for turning down a research opportunity.

What if I'm not offered the position but I really wanted it?

Rejection stings—and it's intensified if you were inspired by the research topic or the interviewer's passion for their research. However, **being turned down for a research position doesn't automatically mean that you weren't qualified, enthusiastic, or professional.** These are possibilities to consider if you're continually invited to interviews but not offered a research opportunity. But for now, let's focus on some of the many reasons an undergrad might not receive an offer to join a research group after an interview that doesn't include a lapse in professionalism at the interview.

Maybe the project was put on hold, for example, or the mentor's life changed so they no longer have time to train a new student. In these cases, an interviewer might inform you of the situation (ideally) or they might feel embarrassed and instead email a "rejection." Other possibilities could be that another student demonstrated stronger genuine interest in the project, their proposed schedule overlapped better with the interviewer's work schedule, or the student previously completed a lab or lecture class that introduced them to the fundamental techniques used by the research group. Some SURPs will turn down applicants who submit lackluster essays (another reason to ensure that you ask a professor, a mentor, an academic advisor, or someone at the career center for help editing them) or those with generic recommendation letters (therefore you should always ask potential references if they can write a strong or enthusiastic letter). We do need mention, however, that some interviewers don't offer a research position to a student because of racism, sexism, ableism, ageism, transphobia, or another unacceptable reason. If you've experienced any of these attitudes, you already know that they are beyond your control, but nevertheless it's infuriating and contributes to the inequity in STEMM.

Some interviewers will inform a student that they didn't get the position at the interview. Even though the news is disappointing, the timing is good because it's immediate feedback, and typically the interviewer gives a reason such as a scheduling issue, overcommitment concerns, or that a student's goal isn't achievable through the available research experience.

If this scenario happens you to and you believe that the interviewer arrived at their decision because you accidentally misstated something, it's appropriate to tell them while you're still at the interview. If they wanted to offer you the position but were concerned because of one of your answers, they will be glad for the clarification and then might ask you to join the project. If, however, after you address the misconception the interviewer still believes that the issue will prevent you from being successful with the available research project, they won't change their mind. Even if this is the case, resist the urge to argue because if you do, it won't persuade the interviewer to reconsider and it's likely they won't recommend you to a labmate or colleague later.

Other interviewers prefer to inform students through email with a variation of, "We interviewed an abundance of qualified candidates for only one position." You have no reason to doubt the sincerity of this statement—after all, you did receive an interview. PIs and other mentors don't conduct research interviews for fun or because they are lonely or bored. They interview students whom they believe could be an asset to the research team. However, it might be helpful to email the interviewer and ask for a specific reason that you were not selected for the position, although there is no guarantee they will respond.

Consider the Interviewer's Feedback If Any Is Given

Whether it happens in the interview room or through email later, if an interviewer shares why they aren't offering you a research position, it's important to consider their reason with an open mind. **You don't have to agree—after all, it's only their opinion—but you should evaluate the feedback to determine whether it seems valid and, if so, how you can improve, if that is an option.** If, for example, you propose five hours per week for research because that is all you have available but nine is the minimum needed to conduct the techniques, there isn't much you can do about it. But one method to strengthen your application or improve interview skills is by learning from those who are more experienced. Even if you believe the interviewer is incorrect, it's helpful to understand their reason because you'll need to change something to make certain the next interviewer doesn't reach the same erroneous conclusion.

If you email an interviewer to ask how you might improve your application for a future position or, specifically why you weren't selected, you might not receive a response. This could be for several reasons, but common ones are an overflowing email inbox, the fear of hurting your feelings, or the concern that responding will be interpreted as an invitation to start a debate on why you believe they made the wrong decision. If the person does respond to your email, even if you don't agree with their suggestions, acknowledge their effort. Reply back with a short (one to three sentence) email thanking them for their advice, following the email format we've already provided.

Reflect on the Interview

Whether or not you receive feedback from an interviewer, reflect on each interview to determine what went well and whether there were any parts that could have gone better. This is good practice after any interview, such as an officer position in a student club, a shadowing position, or for a job or internship. You can enlarge and improve your professional skill set not only by going on interviews but also by reflecting on them considering the following:

- **Did I make several significant interview mistakes?** For example, did the interviewer ask me the same question, or very similar questions, several times? This could mean you didn't answer a question to the interviewer's satisfaction, or they were concerned that you might be misrepresenting something. Again, remember that most people make at least a few small mistakes at an interview and that rarely is the reason they don't get an offer. But you're using self-reflection to consider whether you might have made any of the bigger ones that might have complicated the interview (as described in the section that opened this chapter, "Ten Topics to Help Avoid Interview Mistakes"). Yet don't search for a reason to attack your self-esteem, either. Sometimes you might not receive an offer to join the research project for a reason that has nothing to do with your interview style (as discussed in the previous subsections "What if I'm not offered the position but I really wanted it?" and "Consider the Reviewer's Feedback if Any Is Given").
- **Did the interview end abruptly after I answered a question?** Or did the interviewer seem surprised by an answer but didn't ask me for additional information? Reflect on what was asked and the answer you gave. That might be the key to a statement that gave the interviewer the wrong impression. If you know this, you can avoid a similar situation in your next interview.

Finally, Send a Thank-You Email Even if You Aren't Asked to Join the Research Group

If you don't receive an offer to join the research group, you should still email the interviewer and thank them for their time. If true, mention that you're still interested in working with them on a project in the future if an opportunity becomes available. Also be sure to mention a specific reason that you're interested in their research program that was brought up in the interview. (If you're not interested in the possibility of joining the research group in the future, only thank them for their time.)

Most interviewers won't notice if you don't send a thank-you email, but they will notice (and might be impressed) if you do choose to send a short one. Although sending one probably won't turn a no into a yes, there are other benefits for doing so. Send a thank-you email the same day as your interview regardless of whether you've been offered a position. But don't worry about the time commitment involved because doing this task won't require much time or effort.

Three reasons to send a thank-you email after an interview even if you aren't offered the position:

1. **It's a good habit to develop.** Thanking someone for their time after an interview is a good professional habit to develop. (You'll probably want to follow up with a note of appreciation after many job interviews in your professional life. This action is also a type of hidden "curriculum" that can be expected in smaller companies outside academia.) If sending a thank-you email matters to the interviewer, you win. If it doesn't, sending one shouldn't annoy them, although there is probably someone who is an exception to this (but don't worry about that—send the email).
2. **It helps avoid future awkwardness.** Although it's highly unlikely an *interviewer* will be unprofessional if you bump into each other on campus (or they end up teaching a class you need for your major), sending a thank-you email might make *you* feel less awkward when you meet again. This email is an opportunity to tell the interviewer that you don't have hard feelings about not being invited to join the research group (even if you kind of do for a while), and it might help you work through the disappointment. This will make you feel more comfortable in future interactions in office hours or if the research group you end up joining has cogroup meetings with the one that didn't offer you a position, or the grad student who interviewed you teaches your next lab class.
3. **It makes you more memorable for potential future opportunities.** You might not have been the right candidate for that project, but you could be perfect for the next one. We've interviewed enthusiastic undergrads who had an incompatible course schedule or whose goals weren't obtainable

through the current research position available in our lab. It was easy to recommend these students to colleagues and remember them when we did have a suitable project. Sending a thank-you email gives you a final chance to remind the interviewer how you would be a good addition to their team in the future or their colleague's team now.

EMAIL TEMPLATES TO USE AFTER AN INTERVIEW

Next are email templates to customize in response to several possible interview outcomes.

Email Template to Thank the Interviewer for Their Time (before Decision Is Known)

Dear [Dr. or Person Who Interviewed Me],

I enjoyed meeting with you today and learning about your research program and the available project with your research group. I hope to work with you on the project we discussed.

Thank you for the time you spent in the interview.

Sincerely,
[Your name]
[Your email signature]

* * *

Email Template to Accept the Position

Dear [Dr. or Person Who Interviewed Me],

I'm excited to accept the research position with your research group and can't wait to get started. As agreed, I'll be in the lab 15 hours per week, from 1:00 to 5:00 p.m. on Mon, Tues, Wed, and 1:00 to 3:00 p.m. on Fri.

Is there anything that you suggest I read or learn to prepare for my first day?

Sincerely,
[Your name]
[Your email signature]

* * *

Email Template to Decline an Offer

Dear [Dr. or Person Who Interviewed Me],

I enjoyed meeting with you and discussing the available research project in your lab. After thinking it over, I have decided that it's not the right opportunity for me.

Thank you for the time you spent in the interview.

Sincerely,
[Your name]
[Your email signature]

* * *

Email Template to Ask for Feedback

Dear [Dr. or Person Who Interviewed Me],

I enjoyed discussing your research program even though I am disappointed that I won't be joining your lab as a researcher. Do you have any advice on how I could help prepare for future interviews or strengthen my application to be more competitive for other positions?

Thank you for the time you spent in the interview.

Sincerely,
[Your name]
[Your email signature]

* * *

Email Template for Deal Breakers (Use Only after You've Been Offered the Position)

Dear [Dr. or Person Who Interviewed Me],

I am so excited about this research project because I like the idea of [working on a specific line of research or using techniques discussed at the interview]. However, I have only [*X*] hours to dedicate to research per week—any more and I'll risk becoming overcommitted. If [*Y*] hours per week is firm, I will, unfortunately, need to decline your offer to join the lab.

Sincerely,
[Your name]
[Your email signature]

* * *

CONGRATULATIONS, YOU'RE NOW AN UNDERGRAD IN THE LAB!

The start of a new research experience is equal parts excitement and nervousness. To get the most out of undergrad research, you'll need to do more than show up and put in the hours. You'll need to make the effort to learn about what's going on around you and embrace that idea that your biggest intellectual growth might come not from understanding a result but from thinking of the next question to ask. In other words, to get the most out of your research experience, you'll need to invest yourself in the commitment.

How will you do that? **Make your research experience a priority. Show up on time, ready to work, learn, and contribute.** When you're able, do a little more than the required minimum and adopt "What more can I learn?" as a personal philosophy. **And approach the professional relationships with your in-lab mentor, labmates, and PI as an active participant.** By choosing this approach, you'll gain the greatest benefits and be rewarded with the most fulfilling and meaningful use of your time.

Next are tips to help you prepare for your first day as an undergrad researcher. If you interviewed several weeks before the start date of your position, some of these will be easier to cross off your to-do list than if you interviewed and started on the same day or the following week.

TEN TIPS TO PREPARE FOR YOUR FIRST DAY OF YOUR UNDERGRAD RESEARCH EXPERIENCE

1. **Ask your in-lab mentor what you can do or read to prepare for the first day.** Whether you met in person at the interview or were introduced to them through email, ask your in-lab mentor what you can do to prepare for your first day. They might direct you to do several tasks we list here, or nothing at all. However, if you ask your mentor how to prepare for your first day, it should be a sincere request and you should do your absolute best to follow through with their suggestions. So, even if your mentor gives you a

complicated peer-reviewed article to read, do your best to read it using the chapter 4 section "Tips on Selecting and Reading a Scientific Paper before Applying for a Research Position."

2. **Register for academic credit and complete a research contract.** If you're planning to register for research credit, or a 0-credit option, you might need your PI's permission in writing. Before that can happen, you both might need to complete a research contract. (These contracts can be required even if you're primarily working with a grad student or postdoc in-lab mentor.) The contracts usually cover how many credit hours you'll register for, your key responsibilities, some project details, the PI's basic expectations, and possibly a summary of your PI's mentoring responsibilities. Reviewing this contract before starting is also an insurance policy of sorts because if you're filling out the contract and suddenly realize that the requirements aren't compatible with your academic and life obligations, it's relatively easy to retroactively turn down the position.
3. **Write the research preproposal, if applicable.** Sometimes a student is required to write a preproposal detailing their research project before they are allowed to get started on a project. If this describes your situation, ask your in-lab mentor for advice on getting started. At the very least, create an outline before your first day and visit the writing center on your campus (either in person or virtually) for help. Also, review chapter 3 for additional tips that might be helpful when completing this task.
4. **Complete safety training, if required.** The safety training for your research position will depend on the type of research experience and project you'll be working on. If you're able to take online safety training courses or seminars before starting, do it. (Ask your in-lab mentor about the safety training requirements for your experience if they haven't already volunteered this information.) Vaccinations or classes in animal handling might also be required. But don't consider these classes or seminars as the end of your safety training because they aren't. After starting your research experience in a wet lab, if you're ever unsure of how to do a procedure, use a piece of equipment, or dispose of a chemical or other item, ask a labmate for guidance. Never guess. If you're doing fieldwork, always confirm what you need to bring with you (how much water, sunscreen, gloves, sunglasses, for example), what type of clothing to wear before heading out to a field site, and what precautions to take if you'll have limited cell phone reception or limited access to emergency medical services. It's always better to ask than to guess. Always prioritize safety.
5. **Learn about the techniques you'll use.** If you know the name of a research technique you'll be doing, stream an online video demonstration a few times before your first day. Your goal isn't to memorize the steps of a procedure or even become an expert but to become familiar with the pro-

cedure's language and observe how it's done once or twice. This will make it easier when you start your research project even though your mentor might teach you to do the technique a slightly different way.

6. **Learn about the key words or terms associated with your project.** If your interviewer mentioned key words or terms associated with your project, familiarize yourself with them. Again, your goal isn't to become an expert or memorize definitions but to acquire a basic knowledge of key words and concepts. When you start your research experience, you'll be overloaded with new information, so being comfortable with some project terms ahead of time will be helpful as you tackle learning the language of the project.
7. **Share your new adventure on social media and start connecting with other researchers.** Joining a research group is a big deal and a reason to be proud! Begin your journey of practicing science communication by sharing the news on social media. Start a new Twitter or other social media account or update your bio in in your current one to include your new title as an undergrad researcher. Then follow the accounts belonging to the Office of Undergraduate Research, your college or university, your department, and chapters of scientific organizations related to your field. And if your PI manages a lab group account, follow it. In addition, follow groups (or individuals) you want to connect with because they share your identities. One way to find accounts is to send a tweet or other post asking for suggestions such as "What accounts should I follow to connect with _____" and fill in the blank with, for example, "Black in STEM" or "disabled in STEM" or "LGBTQ in STEM," or include your specific research area in the inquiry, such as "women in ecology" or "Native in biochem" or "Latinas in earth science."
8. **Review your mentor's expectations.** To have the smoothest transition as a new group member, the day before you start your research experience, review the expectations outlined by your mentor in the interview. Most mentors won't spend time going over them again but will expect you to know them. (Don't hesitate to ask a mentor for details if you need clarification, but do this sooner than later. They would rather you ask than guess and be mistaken.) Also, make sure you've added your research schedule to your personal calendar.
9. **Maintain your academic and life balance.** It's only through the conscious practice of time management and prioritizing the activities that are important to you that you'll achieve and maintain a solid academic and life balance. In the third week of your research experience, or sooner if needed, revisit your online scheduling app and adjust your activities as needed. Use the approach in the chapter 4 section "Step 1: Schedule and Prioritize Your Time" each semester to keep your priorities in check without becoming overextended. The small effort spent schedul-

ing is worth the big payoff in productivity, stress reduction, and overall happiness.

10. **Connect with us.** Visit UndergradInTheLab.com for tips, tricks, and strategies related to undergrad research and navigating the hidden curriculum in STEMM. Also, follow us on Twitter @YouInTheLab, and on Instagram @UndergradInTheLab. If you're feeling it, tag us in a selfie and declare, "I'm a new #UndergradInTheLab!" Just be sure that you're wearing the correct personal protective equipment if you snap your photo in the lab or at a field site or we won't be able to share it with our followers. Also, we are working on a companion book to this one that covers the undergrad research experience after you get into a lab, and when there is news about the book, we'll announce it in these spaces.

ACKNOWLEDGMENTS

FROM BOTH OF US

We thank all those who were part of making this new edition a reality. Our gratitude to the undergrad research mentors who shared their perspectives and unedited opinions so we could create a comprehensive resource for undergrads in search of a meaningful research experience in STEMM. Our thanks to all the undergrads who we have had the privilege to mentor in our careers so far and thanks to the students who have connected with us on social media seeking advice or for a place to share their successes. Our endless appreciation to Dan Purich, professor of biochemistry and molecular biology at the University of Florida, for continuous discussions on creating student-centered resources. Thank you to Vertigo Moody, professor and chair, Department of Natural Sciences, Santa Fe College, in Gainesville, Florida, for sharing expertise and for informative discussions. A hat tip to Jennifer Seitz, education and training specialist, University of Florida, for directing our attention to helpful resources. We thank Donna Kridelbaugh, communications consultant, for connecting us with students, professional researchers, and professors on social media and elsewhere. Our appreciation to Madeleine Turcotte, who is long finished with med school but continues to give feedback we have come to rely on. Credit must go to the conscientious team at the University of Chicago Press for guiding this edition over the finish line: Mary Laur, executive editor, for taking the lead and providing expertise; Mollie McFee, senior editorial associate, for completing all the editorial and administrative tasks that we know of (and the ones we were unaware of); Lori Meek Schuldt, the press's contracted copyeditor, for fixing the problems we created; Caterina MacLean, senior production editor, for giving us as much lead time as possible; Kevin Quach, designer, for designing the cover; and Nick Lilly, promotions manager, for helping our readership find their way to us for this book. And thank you to the anonymous reviewers who generously traded their valuable time (and we suspect substantial portions of an academic break) for suggestions that helped us create a better manuscript.

FROM PARIS

Thank you to Brandi O., who doesn't care if I show up wearing pajamas; to Mike S., whose stories I would not believe if they came from someone else; to Susan T., for more reasons than I could share here; and to Mary B., who responds lovingly to every single story I share about my rabbit, Snickers.

FROM DAVID

Cheers to all the members of The Running Tabs for the camaraderie.

INDEX